the
UNIVERSITY
of
GREENWICH

DEVELOPMENTS IN AGRICULTURAL
AND MANAGED–FOREST ECOLOGY 22

tropical forests

DEVELOPMENTS IN AGRICULTURAL
AND MANAGED–FOREST ECOLOGY 22

tropical forests

some African and Asian case studies of composition and structure

JÁN BOROTA

University College of Forestry and Wood Technology
Zvolen, Czechoslovakia

ELSEVIER
Amsterdam – Oxford – New York – Tokyo 1991

Scientific Editor
Academician Adolf Priesol

Published in co-edition with
VEDA, Publishing House of the Slovak Academy of Sciences, Bratislava, Czechoslovakia

Distribution of this book is being handled by the following publishers:

for the U.S.A. and Canada
ELSEVIER SCIENCE PUBLISHING COMPANY, INC.
655 Avenue of Americas
New York, NY 10010

for the East European countries, China, Cuba, Korea, Mongolia and Vietnam
VEDA, Publishing House of the Slovak Academy of Sciences
Klemensova 19
814 30 Bratislava, Czechoslovakia

for all remaining areas
ELSEVIER SCIENCE PUBLISHERS
25 Sara Burgerhartstraat
P. O. Box 211, 1000 AE Amsterdam, The Netherlands

Library of Congress Cataloging-in-Publication Data

Borota, Ján, 1925–
    Tropical forests / Ján Borota ; [translation František Papánek].
       p.   cm. -- (Developments in agricultural ana managed-forest
  ecology ; 22)
    Translated from the Slovak manuscript.
    "Published in co-edition with VEDA, Bratislava, Czechoslovakia."
    Includes bibliographical references and index.
    ISBN 0-444-98768-1 (Elsevier)
    1. Forests and forestry--Tropics. 2. Rain forests. I. Title.
  II. Series.
    SD247.B67 1990
    634.9'0913--dc20   90-44114
                                           CIP

ISBN 0-444-98768-1 (Vol. 22) (Elsevier)
ISBN 0-444-41515-7 (Series)
ISBN 80-224-0082-3 (Veda)

Printed in Czechoslovakia.

OTHER TITLES IN THIS SERIES

Preface

This publication deals with solving the problems concerning the composition and structure of both natural and artificially established tropical forests. It is a self-contained synthesis of the author's published papers and of unpublished work representing the results of twenty years' study and consultative activities in various countries of Asia and Africa.

Tropical forest communities occur mainly on the territory between the Tropic of Cancer and the Tropic of Capricorn. In this region which has a surface area of about 52 million square kilometers, almost 1,600 million people (i.e. about one third of the Earth's population), are living. The natural vegetation which covers most of the tropical region is included within a number of tropical ecosystems.

Together with the growing human population in developing countries the consumption of wood, especially that of fuelwood, is also on the increase. Most developing countries depend on forest resources as their main energy supply. In many developing countries the population's sustenance is obtained by expanding arable land to the detriment of the forest. Shifting cultivation which afflicts all types of tropical forest formations in almost all developing countries is the chief cause for the decline and deterioration of tropical forests.

Because of the systematically diminishing surface of tropical forests – according to FAO statistics at present 7.3 million ha of tropical forest disappear annually – the conservation of tropical forests is seen world-wide as being of the utmost urgency for maintaining the original natural environment. For example, in 1977 the World Wildlife Fund (WWF) has urged Bolivia, Brazil, Colombia, Peru and Venezuela to safeguard viable representative samples of the evergreen forests of the Amazonian basin. It recognizes the need to make productive use of the Amazonian area and the conservation efforts.

From the ecological point of view and with regard to biomass production, the evergreen moist tropical forest is the most luxuriant and most complicated vegetational formation in the world. In spite of such luxuriant growth the forest soil is usually deficient in nutrients as well as in other types of forest formations. After deforestation the nutrients of the soil become exhausted in a few years, the yields of agricultural crops diminish, and the land becomes infertile. The conservation of forest resources in developing countries can be achieved by an integrated development of the countryside, and by the improvement of living standards of the rural population in harmony with the development of agriculture and forestry (agroforestry) while respecting ecological viewpoints.

From the point of view of production and commerce information on the composition and structure of tropical forests is of great importance, mainly for planning for the rational utilization of tropical forests in the relevant developing countries. The development of forestry and forest industries in these countries requires more detailed information on timber resources. It also requires that the original forest is rationally protected against deterioration and that it is also protected from excessive deforestation, over-cutting or any impairment of the ecological balance.

Before discussing these problems some information based on the literature, statistical data and the author's own knowledge of the types of world-wide forest formations and information on the importance of forests as a source of timber will be given.

In subsequent chapters the author presents information on the species composition of forests and the diameter structure of trees in natural tropical forests. By analysing inventory data from selected sites in Ghana, Congo, Gabon and Laos, basic information is obtained on the timber resources in these countries, and the present situation and the possibilities for developing their forest management are discussed.

Though there are many publications dealing with the structure of natural tropical forests it should be noted that the actual definition itself of the structure of tropical forests lacks unity (Brüning, 1970; Richards, 1952; Lamprecht, 1972; Rollet, 1974). From a practical point of view it seems to be most important that the diameter structure of trees in tropical forests is considered, i.e. the frequency distribution of trees by their diameter breast height for a given area.

The analysis of inventory data was based on the diameter distribution of trees according to the various tree species or their groups; this being an important indicator of the structure and yield of timber assortments in the natural tropical forest. Data obtained in this way provide detailed information on the species composition, diameter structure of trees, and on timber production in both evergreen and semi-deciduous moist tropical forests in the developing countries mentioned.

Interesting results were obtained by the author when studying problems concerned with dry deciduous tropical forests of the "miombo" type in Tanzania. As a consequence of the forest fires which cause damage to the ecosystem almost regularly each year, a smaller number of trees were found to have been established in the smaller diameter classes. In these forests, trees of economically important species are characterized by a small annual diameter and volume increment.

Special attention in this book will be paid to the growth processes found in forest plantations. Using certain tropical and subtropical tree species which have been widely introduced and cultivated in many developing countries, dendrometrical measurements on permanent experimental plots were evaluated and the diameter and height structure of trees and their increments have been analysed.

A separate chapter is devoted to thinnings in artificially established stands of *Pinus patula*. Here, results of measurements on permanent experimental plots established in Tanzania, mainly in the region of mountain savannas, are analysed.

Although the primary function of forest plantations cultivated in the tropics and subtropics is for the production of timber in a shorter time span than is possible in natural forests, artificially established man-made forests also provide other benefits which may promote the conservation and protection of the natural environment.

Problems of tree species composition and diameter structure of trees in both natural and artificially established tropical forests were studied by the author in various ecological conditions, where temperature and the amount and distribution of precipitation are the main climatic factors. An explanation is given in this publication about the narrow connection between climatic factors and timber production – the main function of the forest – and the forest's useful social and environmental influences.

The book is intended for scientific and professional workers, college teachers and students, organizations and, in general, all those interested in problems concerning the tropical forest and forestry, including the commerce and processing of tropical timber.

Ján Borota

Contents

1 Methodology of work

The methodological approach to the processing of empirical data was preceded by the study and evaluation of scientific literature including statistical data published by international agencies of the United Nations (FAO, UNESCO, ECE) and other unpublished material. In this way basic information was obtained about the natural and economic conditions, the types of forest formations, and the situation and distribution of timber resources in the developing countries investigated.

The rational utilization of tropical natural forests requires some basic information about their tree species composition and especially about the diameter structure of trees. For this aim, the author used figures from forest inventories made on selected sites in Ghana, Congo, Gabon and Laos. Inventory methods used in these countries differed and they are described in Chapter 4 (4.1.4, 4.2.3, 4.3.3 and 4.4.3) in conjunction with the solution of problems in the countries mentioned. The planning of forest management in natural tropical forests has been approached only in connection with an inventory of the Sibiti-Zanaga and Kouilou forest regions in the Republic of Congo. This methodology is described in Chapter 4 (4.2.4 and 4.2.5).

Using examples from the developing countries mentioned, the types of tropical forests and their tree species composition were analysed with regard to the economic importance of the various timber species. The occurrence of *Aucoumea klaineana* in Congo and Gabon was studied in more detail.

The author used the beta function formula for the theoretical expression of tree distribution by diameter breast height, and height, respectively, in natural tropical forests.

The tree species composition and diameter structure of trees in dry deciduous tropical forests of the miombo type were analysed using statistical methods for the evaluation of data obtained by measurements made on permanent experimental plots. Detailed analysis has been made on the Sikitiko permanent experimental plot where the economically important species of *Pterocarpus angolensis* was abundant.

The author also used statistical methods for the analysis of the diameter and height distribution of trees in forest plantations for the analysis of increment and timber production of *Tectona grandis, Shorea robusta, Eucalyptus globulus, E. regnans, Cedrus deodara, Juniperus procera, Pinus caribaea* and *P. patula*.

For all these species the measured values of diameter breast height and tree height obtained on permanent experimental plots are analysed. As to *Tectona grandis* and *Shorea robusta* the data were also analysed according to yield tables used in India. Data from forest inventories and thinning plots were processed and evaluated in the same way using statistical methods.

The χ^2 test of good agreement was used for testing the calculated values. In the analysis of natural tropical forests the Kolmogorov–Smirnov test was used for the verification of calculated values.

2 Types of forest formations throughout the world

Forests including woodland savanna and bush cover about one third of dry land area of our planet. The surface of closed forests which are or could be used for timber production is about 2,950 million ha, which is about 22% of the dry land surface. Of this, broad-leaved forests represent about 62% and coniferous forests 38%.

The climate exercises a marked influence on the vegetation of the globe, shaping ecologically defined geographical forest units – types of forest formations which include lower units, various forest types (Weck, 1957) and biomes.

The occurrence of various types of forest formations is conditioned by important climatic factors such as temperature, precipitation, and their mutual relationships. In tropical conditions the duration of the rainy and dry seasons together with the amount and distribution of annual precipitation, as well as other factors, especially those of the soil and altitude are important conditions. Various authors, including both natural historians and foresters have proposed classifications of vegetation or types of forest formations according to the climate (Schimper, 1903; Mayr, 1909; Champion, 1936; Svoboda, 1952; Sasson et al., 1978).

Table 1. Approximate areas of the world's forests

Forest formation	Area in millions of hectares
Temperate zone	
Boreal coniferous forests	700
Mixed forests	500
Subtropical zone	
Moist forests	30
Dry forests	200
Tropical zone	
Moist evergreen forests	580
Moist deciduous forests	320
Dry deciduous forests	620
Total	2,950

For a general orientation and in agreement with statistical data published by international agencies of the United Nations (Economic Commission for Europe – ECE, Food and Agricultural Organization – FAO, United Nations Educational, Scientific and Cultural Organization – UNESCO, United Nations Environmental Programme – UNEP) the vast forest masses of the world can be classified into the formations given in Table 1.

The main types of world forest formations are schematically shown in Fig. 1.

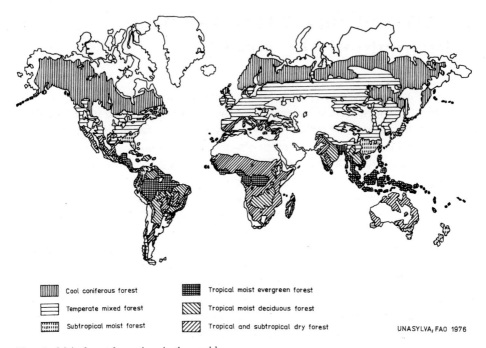

Cool coniferous forest	Tropical moist evergreen forest
Temperate mixed forest	Tropical moist deciduous forest
Subtropical moist forest	Tropical and subtropical dry forest

UNASYLVA, FAO 1976

Fig. 1. Main forest formations in the world.

2.1 Forests of the temperate zone

Boreal coniferous forests, also called taiga, extend over the Northern Hemisphere in the temperate zone of Europe, Asia and America approximately between 50 and 72° of northern latitude. They occur in the cooler zone with a growing period of 60–120 days, an average annual temperature below 10°C and an annual precipitation of 600–1,000 mm or more.

Mixed forests of the temperate zone also occur in the Northern Hemisphere in Europe. Asia and America approximately between 38 and 50° N. The growing period lasts from 100 to 150 days, the average annual precipitation is between 600 to 1,200 mm and in some locations it is even more.

Whereas the coniferous boreal forests are rather homogeneous, with a small number of tree species and a poor growing stock (on average about 60–100 m³ per hectare) (Weck and Wiebecke, 1961), in the mixed forests of the temperate zone there is a larger number of both broad-leaved and coniferous tree species, and the average growing stock in this forest formation ranges from 150 to 300 m³ per hectare.

European boreal coniferous forests occur in Scandinavia and in northern regions of the U.S.S.R. the north of the 58th parallel. The most important coniferous economic tree species of these forests are *Pinus sylvestris* L. and *Picea abies* (L.) KARST.; for broad-leaved species *Betula pubescens* EHRH., *B. pendula* ROTH and *Populus tremula* L.

In Asia, Siberia and the Far East, the most important coniferous economic tree species of the taiga are the larch species *Larix sibirica* (MUENCHH.) LEDEB. and *L. gmelini* (RUPR.) KUZENEVA; the pines *Pinus sylvestris* L., *P. sibirica* DU TOUR, *P. koraiensis* SIEB. et ZUCC. and the shrubby *P. pumila* REGEL; the spruce species *Picea obovata* LEDEB. and *P. orientalis* (L.) LINK; and the fir *Abies sibirica* LEDEB. From among over forty birch species *Betula pubescens* EHRH., *B. pendula* ROTH, and *B. dahurica* PALL. – as well as aspen *Populus tremula* L. – are the most important.

The American boreal coniferous forests occur in the northeast and northwest of the U.S.A. and in the eastern, central and western regions of Canada. They are characterized by a larger representation of economic tree species than are the boreal forests of Europe and Asia (Haden-Guest, 1956).

Among the main economic coniferous tree species of boreal forests in the eastern and central regions of Canada and in the northeastern regions of the U.S.A. are *Abies balsamea* (L.) MILL., *Larix laricina* (DU ROI) K. KOCH., *Picea glauca* (MOENCH) VOSS, *P. mariana* (MILL.) B. S. P., *P. rubens* SARG., *Pinus banksiana* LAMB., *P. strobus* L., *P. resinosa* AIT., *Thuja occidentalis* L. and *Tsuga canadensis* (L.) CARR. The most important broad-leaved tree species are *Betula lenta* L., *B. lutea* MICHX., *B. papyrifera* MARSH., *Acer rubrum* L., *A. negundo* L., *Fagus grandifolia* EHRH., *Fraxinus nigra* MARSH., *Populus balsamifera* L., *P. tremuloides* MICHX., *Quercus borealis* MICHX. f., *Tilia americana* L. and *Ulmus americana* L.

In the northwestern part of the U.S.A. and in the western and northwestern regions of Canada economically important coniferous tree species are represented by *Abies grandis* (DOUGL. ex D. DON) LINDL., *A. lasiocarpa* (HOOK.) NUTT., *Larix occidentalis* NUTT., *Picea engelmannii* PARRY ex ENGELM., *P. sitchensis* (BONG.) CARR., *Pinus contorta* DOUGL., *P. monticola* DOUGL., *P. ponderosa* LAWS., *Pseudotsuga menziesii* (MIRB.) FRANCO, *Thuja plicata* DONN and *Tsuga heterophylla* (RAF.) SARG. Economically important broad-leaved tree species of the western region of boreal forests include *Acer macrophyllum* PURSH., *A. negundo* L., *Betula papyrifera* MARSH., *Fraxinus oregona* NUTT., *Populus tremuloides* MICHX., *P. trichocarpa* TORR. et GRAY and *Quercus garryana* DOUGL. (Stefferud et al., 1949).

In the mixed forests of the temperate zone of Europe important economic tree species include *Abies alba* MILL., *Larix decidua* MILL., *Picea abies* (L.) KARST.,

Fig. 2. Mixed forest of the temperate zone with *Fagus silvatica* and *Picea abies* as the main tree species. Region of Zvolen, Czechoslovakia.

Fig. 3. Border natural regeneration of a beech stand. Voderady Forest Reserve, Czechoslovakia.

▶

Fig. 4. The tallest coniferous tree species of the world; *Sequoiadendron giganteum*. Josemite National Park, California, U.S.A. (Photo: L. Borotová)

Pinus sylvestris L., *Acer pseudoplatanus* L., *A. platanoides* L., *Fagus silvatica* L., *Fraxinus excelsior* L., *Quercus petraea* (MATTUSCH.) LIEBL., *Q. pubescens* WILLD., *Q. robur* L. and *Ulmus glabra* HUDS. (Figs 2, 3).

In the Asian mixed forests of the temperate zone the following economically important tree species occur: *Pinus sylvestris* var. *mongolica* LITVINOV, *Larix gmelini* (RUPR.) KUZENEVA, *Betula platyphylla* SUKATSCH., *Acer ginnala* MAXIM., *Castanea mollissima* BLUME, *Quercus dentata* THUNB., and other tree species (Wangh, 1961).

Economically important coniferous tree species growing in mixed forests of the temperate zone in the eastern part of North America include *Juniperus virginiana* L., *Pinus rigida* MILL., *P. strobus* L. and *Taxodium distichum* (L.) RICH. Of main economic importance among broad-leaved tree species are mainly *Quercus alba* L., *Q. bicolor* WILLD. and *Q. velutina* LAMK., as well as *Acer saccharinum* L., *A. saccharum* MARSH., *Carya cordiformis* (WANGENH.) K. KOCH, *C. tomentosa* NUTT., *Diospyros virginiana* L., *Fagus grandifolia* EHRH., *Fraxinus americana* L., *Juglans nigra* L., *Liquidambar styraciflua* L., *Platanus occidentalis* L., *Populus deltoides* BARTR., *Ulmus americana* L. and *U. rubra* MUEHLENB.

Coniferous tree species which occur most frequently in the western regions of mixed forests of the temperate zone in North America are *Abies concolor* (GORD. et GLEND.) HOOPES, *Libocedrus decurrens* TORR., *Picea pungens* ENGELM., *Pinus lambertiana* DOUGL. and *Pseudotsuga menziesii* (MIRB.) FRANCO. In the Rocky Mountains the tallest coniferous tree species of the world are found, these are *Sequoia sempervirens* ENDL. and *Sequoiadendron giganteum* (LINDL.) BUCHH. (Fig. 4). Broad-leaved economic tree species in this region include *Alnus rubra* BONG., *Castanopsis chrysophylla* (DOUGL.) A. DC., *Populus sargentii* DODE, *Quercus californica* (TORR.) COOPER, *Q. garryana* DOUGL. and others.

The boreal coniferous and the mixed forests of the temperate zone are of great importance to the forest industries of economically developed countries. The annual production (fellings) of rough timber in these types of forest formations is estimated to be over 1,200 million m^3, which is more then 40 % of the world's timber production, three quarters of it being represented by coniferous timber.

2.2 Forests of the subtropical zone

Subtropical forests occur in both hemispheres approximately between the tropics and the 38th parallel (in Tasmania they extend as far as the 43th southern parallel).

Moist subtropical forests resemble moist evergreen tropical forests by their appearance, tree structure and their floristically rich undergrowth. They occur where the average annual precipitation exceeds 1000 mm and where the average annual temperature is above 10°C. They grow at altitudes from lowlands to 1,200 m and more.

Important tree species of the moist subtropical forests in eastern Asia include, for example, from among coniferous species: *Cryptomeria japonica* D. DON., *Cunninghamia lanceolata (LAMB.) HOOK., *Pinus densiflora* SIEB. et ZUCC., *P. massoniana* LAMB., and species of the *Juniperus* L. and *Pseudolarix* GORD. genera; and from among broad-leaved tree species there occur mainly *Cinnamomum camphora* (L.) SIEB. and species of the *Castanopsis* SPACH., *Michelia* L., and *Schima* REINW. genera.

In the moist subtropical forests of Australia eucalypts are the most economically important species, for example, *Eucalyptus globulus* LABILL., *E. deglupta* BL., *E. delegatensis* R. T. BAK. and *E. regnans* F. MUELL. which attain heights of 70 to 90 m. Important among coniferous tree species are *Araucaria cunninghamii* SWEET and *Agathis robusta* F. M. BAILEY; and in New Zealand broad-leaved species of the *Nothofagus* BL. genus, and coniferous species of *Dacrydium cupressinum* SOLAND. and *Agathis australis* SALISB. are important (Brown et al., 1968).

In South America moist subtropical forests occur in the central parts of Chile, eastern Paraguay, Argentina and southern Brazil. Among the coniferous species of *Araucaria angustifolia* (BERT.) O. KUNZE and *A. araucana* (MOLINAR) K. KOCH are economically the most important as are mainly species of the *Nothofagus* BL. genus among hardwoods.

The moist subtropical forest formation also includes the southeastern and southern parts of the U.S.A. The important conifers here include: *Pinus elliottii* ENGELM., *P. palustris* MILL., *P. taeda* L., *Juniperus virginiana* L. and *Taxodium distichum* (L.) RICH.; and important broad-leaved species include: *Quercus stellata* WANGENH., *Q. falcata* MICHX., *Carya glabra* (MILL.) SWEET and others. In Mexico, pines of the moist subtropical mountain forests which total 38 species are important. *Pinus patula* SCHLECHTD. et CHAM., *P. ayacahuite* ERENB., *P. leiophylla* SCHLECHTD. et CHAM., *P. montezumae* LAMB., *P. oocarpa* SCHIEDE, *P. pseudostrobus* LINDL., *P. tenuifolia* SHAW., and also *Abies religiosa* SCHLECHTD. et CHAM. and *Cupressus lusitanica* MILL. grow at altitudes of 1,500–3,000 m where there is an annual precipitation from 600 to 1,000 mm.

Subtropical dry forests, also called Etesian forests, are of low growth, and from the point of view of timber production are of less importance. They occur where annual precipitation exceeds 500 mm and the mean temperature is above 10°C.

In the Mediterranean the subtropical forests consist mainly of the following tree species: *Quercus ilex* L., *Q. suber* L., *Q. coccifera* L., *Olea europaea* L., *Pistacia lentiscus* L., *Pinus pinaster* AIT., *P. pinea* L., *P. halepensis* MILL., *P. brutia* TEN., *Cupressus sempervirens* L. and species of the *Juniperus* L. genus. The trees attain heights of 5–15 m (Fig. 5).

In southern Africa there are dry, floristically rich subtropical forests, called "fynbosch" or "macchia". The trees attain heights of 3–10 m.

From Afghanistan to India and Burma there are subtropical forests at altitudes of 900–2,000 m. These mountain forests occur where there is a mean annual temperature of 15–18°C and where the annual precipitation is 1,000 mm and more. Economically the most important broad-leaved tree species are: *Quercus incana* ROXB., *Q. semecarpifolia* SM., *Aesculus indica* COLEBR. and *Schima wallichii*

Fig. 6. Temperate-montane fir-spruce forests in the Himalayas. *Pinus wallichiana* in the foreground. Kulu Forest Division, Punjab, India.

Fig. 7. *Abies pindrow* trees, height 55–60 m. Altitude 2,500 m. Manali Forest Reserve, Kulu Forest Division, Punjab, India.

◀

Fig. 5. Subtropical Mediterranean forest with *Pinus halepensis, Cupressus sempervirens* and species of the *Quercus* genus as the main tree species. Region of Orebić, Yugoslavia.

11

CHOISY; and among conifers, *Pinus roxburghii* SARG. and *P. kesiya* ROYLE are the most important.

In Australia various eucalypts and acacias are the main tree species of dry subtropical forests, for example *Eucalyptus bicostata* MAID., *E. gomphocephala* DC. and *Acacia mearnsii* DE WILD.

In California, Arizona and Mexico dry subtropical forests are called "chaparral". In these open forests of low growth, many oak species, such as *Quercus agrifolia* NÉE and *Q. lobata* NÉE, and other species including species of the *Liquidambar* L., *Euphorbia* L. and *Myrica* L. genera are represented.

The so-called temperate-montane forests can be considered as a special type of forest formation. These mountain forests of the subtropical zone occur in conditions where there is a short period of dormancy in the cool season. Annual precipitation varies between 800 and 1,500 mm or more, and the mean annual temperature does not usually fall below 10°C. These forests occur both in the Northern and Southern Hemispheres.

Temperate-montane forests occur in the Himalayan mountains at altitudes of 2,000–3,300 m. Important tree species grow here, such as *Cedrus deodara* (ROXB.) LOUD., *Abies pindrow* ROYLE, *Picea smithiana* BOISS., *Pinus wallichiana* JACHS. (Fig. 6), *Acer caesium* WALL. and *Aesculus indica* COLEBR. The trees in closed stands attain heights of 10–40 m and on fertile soil even up to 50 to 60 m (e.g. *Abies pindrow* ROYLE) (Fig. 7).

2.3 Tropical forests

Tropical forests are conditioned, as a rule, by the climate, and are only exceptionally conditioned edaphically (e.g. mangrove forests, tropical forest communities on swamps). Climatically conditioned forest types depend mainly on the amount and distribution of the annual precipitation. Precipitation is reflected in the type of the tropical forest, especially in its greater or smaller humidity or dryness. The most pronounced characteristic of the influence of precipitation on tropical forests is on the deciduousness of leaves. In this respect, there are continuous transitions from evergreen forest to mostly evergreen, semi-deciduous, mostly deciduous and deciduous forest.

Tropical forests represent more than 50% of the area of all forests throughout the world. They extend in the tropical belt of Africa, Asia, Oceania and Latin America both to the north and south of the equator approximately to the tropics of Cancer and Capricorn. They occur on lowlands, hills and mountains which have varied amounts and different distributions of precipitation over the year, and they are usually found where there is a high temperature which does not sink below freezing point.

2.3.1 Evergreen moist tropical forests

Evergreen moist tropical forests, also called rain forests, grow in regions having a hot moist tropical climate with an almost evenly distributed annual precipitation exceeding 1,400 mm and a high relative air humidity (80–95%). The average annual temperature (20–28° C) is without major daily or monthly fluctuation. Evergreen moist tropical forests occur on various solid, swampy or periodically inundated soils. As to its appearance, the evergreen moist tropical forest remains unchanged throughout the year (Jeník, 1973).

From the ecological point of view and as regards biomass production, the evergreen moist tropical forest is the richest and most complicated vegetational formation in the world. The total plant biomass in a dry state amounts in tropical forests to 400–660 t per hectare, whereas in mature forest stands of the temperate zone it amounts to about 300 t per hectare (Hadley and Lanly, 1983). Evergreen moist tropical forests usually consist of a compact closed tall growth including a large number of various evergreen tree species, trees of different shapes, shrubs, herbaceous and woody lianas (often with a diameter of up to 30 cm and a length of several dozen metres), many epiphytes, herbs, ferns and grasses which may attain heights of up to 5 m, and an abundance of saprophytic, parasitic and other lower plant species (Figs 8–10).

The evergreen moist tropical forest also abounds in animal life which is adapted to the dark, moist and hot forest environment. The animals influence the ecology of the tropical forest, they pollinate flowers, collect, break open and eat the fruits and seeds of trees and by their droppings they distribute seeds over a large area, thus assisting in the germination and growth of various plant species (Baur, 1964).

A special characteristic of the evergreen moist tropical forest is its special diameter and height structure and the spatial distribution of trees. In the evergreen moist tropical forest certain light demanding tree species grow to a height of 50 to 60 m and to a diameter of 150–350 cm, while other tree species attain a height of a few metres and a diameter of a few centimetres only. In the evergreen tropical forest in Borneo the highest tree of the species *Koompasia excelsa* TAUB. was measured 84 m and in the Malay Peninsula 81 m (Whitmore, 1975). A log of *Entandrophragma cylindricum* SPRAGUE in Ghana, in the Pra river region, was 27 m long with a butt end diameter of 226 cm and a top measure of 180 cm; its volume was 88 m^3. In the course of a forest inventory in the Sibiti-Zanaga region in Congo the largest tree dimensions of utile *(Entandrophragma utile)* were measured. The span between buttresses was 14 m, their heights were 16 m, the trunk diameter above the buttresses was 3.50 m, the trunk length from the buttresses to the first branches of the tree crown was 25 m and the height of the tree was about 60 m. The tree crown was broken (Borota, 1984).

Longevity of trees in moist evergreen tropical forest generally does not surpass 400 years but there are exceptions: *Bartholletia excelsa* in the Jari river region, Brazil, had a girth of 14 m and its age might exceed 1,400 years (Pires, 1978).

13

Fig. 8. Evergreen moist tropical forests. Mayombe Mountains, Congo.

The structure of the evergreen moist tropical forest consists of several height strata of trees of varied diameter and of various ages (Davis and Richards, 1933, 1934; Banerji, 1957) (Fig. 11). Schematically, five strata of tree height are distinguished. The highest trees of the upper storey have, as a rule, straight cylindrical boles free from branches with various, sometimes large crowns. The trees of the upper storey rarely form a closed canopy. Large trees of certain species form, mainly on soft ground, buttresses reaching to a height of 2–8 m, e.g. *Piptadeniastrum africanum* BRENAN in Africa. On wet ground certain tree species form support roots, e.g. species of the *Uapaca* BAILL. genus.

Compared to the mountain evergreen tropical forests the lowland forests are floristically more luxuriant and the trees attain larger dimensions. In mountain forests the growth includes more ferns and epiphytes, and less lianas, and the

14

Fig. 9. Vegetation in the lower storey of an evergreen moist tropical forest. Berekum Forest Reserve, Ghana.

Fig. 10. Buttresses of *Terminalia superba* in a semi-deciduous moist tropical forest. N'Gouha II, Congo.

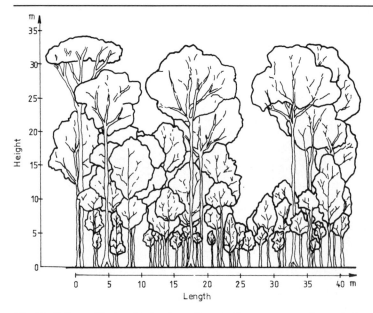

Fig. 11. Pattern of the height structure of trees in the evergreen moist tropical forest.

Fig. 12. Undergrowth of an evergreen tropical mountain forest. Shume Forest Reserve, West Usambara Mountains, Tanzania.

16

number of tree species per given unit area is smaller than in lowland forests (Fig. 12).

Another typical characteristic of tropical evergreen forests is the occurrence of a large number of tree species per surface area unit. In the Sibiti-Zanaga region in Congo which has an area of 940,000 ha, a total of 2,800 ha of half-hectare inventory plots was evaluated. In this area 396 tree species with a diameter breast height of 60 or more cm were recorded. In the Malay Peninsula in Pasoh 5,907 trees belonging to 460 different species were recorded over an area of 11 ha. On the island of Puerto Rico about 550 autochthonous tree species were observed.

The evergreen moist tropical forests in Africa are confined to two main regions. The western region includes a 200–400 km wide coastal belt in Sierra Leone, southern Guinea, Liberia, Ivory Coast and Ghana and is separated by a belt of grass and woodland savanna in the territory of Togo and Benin from the more extensive equatorial region in Cameroon and Congo which reaches as far as 2,000 km from the coast. This region includes the southern part of Nigeria, southern Cameroon, the southernmost part of the Central African Republic, Equatorial Guinea, Gabon, Congo and the northern parts of Angola and Zaire.

The characteristic tree species of the western region of evergreen and semi-deciduous moist tropical forests of Africa are as follows: *Lophira alata* BANK ex GAERTN. f., *Turraeanthus africana* PELLEGR., *Tarrietia utilis* SPRAGUE and species of the *Uapaca* BAILL., *Entandrophragma* C. DC., *Khaya* A. JUSS., and *Celtis* L. genera. Species including *Brachystegia laurentii* LOUIS, *Gilbertiodendron dewevrei* J. LÉONARD, species of the *Celtis* and *Entandrophragma* genera and *Terminalia superba* ENGL. et DIELS. dominate in the central equatorial region and *Aucoumea klaineana* PIERRE dominates in Gabon. The most important commercial tree species of evergreen tropical forests in Africa belong to the Meliaceae family, e.g. the *Entandrophragma, Khaya, Lovoa* HARMS and other genera (Fig. 13).

In Asia evergreen and semi-deciduous moist tropical forests occur in India in the Western Ghats (Fig. 14), in Assam and on the Andaman Islands, as well as in the Chittagong mountains of Bangladesh, in southern Burma, Thailand, Vietnam, Cambodia, Sri Lanka, Malaysia, the Philippines, Indonesia and on the Pacific Islands to the east of Australia.

The most widespread commercial tree species of the evergreen moist tropical forests in Asia belong to the Dipterocarpaceae family, mainly to the *Dipterocarpus* GAERTN. f., *Shorea* ROXB., *Parashorea* KURZ, *Pentacme* A. DC. and *Hopea* ROXB. genera (Cousens, 1965). In tropical Asia 470 tree species included in 12 genera belong to the Dipterocarpaceae family. In the evergreen moist tropical forests of Australia the Myrtaceae family with tree species of the *Eucalyptus* L'HÉRIT. genus is the most important one.

In the evergreen moist tropical mountain forests of Malaysia and Indonesia at altitudes of 2,000–3,500 m there also occur the coniferous tree species of the *Araucaria* JUSS., *Agathis* (LAMB.) STEUD., *Dacrydium* SOLAND., *Podocarpus* PERS. and *Libocedrus* ENDL. genera.

17

Fig. 13. Semi-deciduous moist tropical forest with representation of trees of the Meliaceae family. Sibiti Forest Reserve, Congo.

Fig. 14. Timber conversion of wood assortments by axe in a moist tropical forest. Nilambur Forest Division, India.

Asian evergreen moist tropical forests are considered to be the floristically richest variant of the tropical forest in the world. The average growing stock for tree diameters of 10 cm and more amounts to 130–350 m³ per hectare. The volume of extraction of logs of commercial tree species varies from between 20 and 100 m³ per hectare (Whitemore, 1975).

In Latin America evergreen moist tropical forests also known as "selvas" or "hylaea" are found in the south of Mexico, in the eastern lowland regions of the Central American republics, on the Antilles and mainly in South America in the vast river basins of the Amazonas and Orinoco and in parts of Guyana, Surinam and French Guyana bordering on the Atlantic (Walter, 1962). The area of evergreen moist forests in the Amazonas region amounts to about 340 million hectares which represents the largest compact formation of the evergreen tropical forest in the world. About 2,500 different tree species are to be found growing here. The average growing stock for trees of a diameter breast height of 10 cm and more on various locations varies from 100 to 280 m³ per hectare. The actual extracted log volume of commercial species is around 10–25 m³ per hectare (Pires, 1978).

In the evergreen tropical forests of Latin America an important part of commercial tree species is included in the Meliaceae family (the *Carapa* AUBL., *Cedrela* P. BR.., *Guarea* L. and *Swietenia* JACQ. genera) and in the Leguminosae family (the *Andira* URB., *Caesalpinia* L., *Dicorynia* BENTH. and *Peltogyne* BENTH. genera). Tree species belonging to other families, e.g. *Calophyllum brasiliense* CAMB., *Hura crepitans* L., *Manilkara bidentata* A. CHEV., *Ochroma pyramidale* URB., *Ocotea rodiaei* MEZ, *Pheobe porosa* MEZ, *Virola surinamensis* WARB. and others are also of economic importance (Aubréville, 1961).

A special type of the evergreen tropical forest is the mangrove forest. It occurs in the tropical belt on the sea-shore, mainly at estuaries of large rivers. A special feature of the mangrove forest is its inundation twice daily by mixed salty and fresh water during the rising tide. At low-tide the water level is falling and the trees together with their breathing and support roots are uncovered.

Compared to other tropical forests, the number of tree species in the mangrove forest is much smaller, totalling about 90 species in all tropical regions taken together. Among them, species of the *Avicennia* L., *Rhizophora* L., *Heritiera* AIT., *Bruguiera* LAMB. and *Sonneratia* L. f. genera are more important. The trees attain a height of 8–15 m, as a rule, but may reach in exceptional cases even 25–30 m. The surface of all tropical mangrove forests in the world is estimated to be 20 million ha (Christensen, 1983).

2.3.2 Moist deciduous tropical forests

Moist deciduous tropical forests occur in the tropical zone wherever monsoon rains give rise to precipitation between 1,200 and 2,000 mm. The rainy season lasts for 4 to 6 months, the average annual relative humidity of the air is 60–80% and the

Fig. 15. Experimental plot in a moist deciduous tropical mountain forest with *Ocotea usambarensis* as the main tree species. Southern Kilimanjaro, Tanzania.

mean annual temperature is 20–30° C. In the dry season most tree species shed their foliage.

Moist deciduous tropical forests are composed of several storeys and their tree structure as to height and diameter is differentiated. The tallest trees attain a height of 30–45 m and a diameter of 80–150 cm. A smaller number of tree species per area unit occurs here, as compared to the evergreen moist tropical forests.

In Africa moist deciduous tropical forests occur mainly in the neighbourhood of evergreen moist tropical forests. Important commercial tree species grow here, such as *Antiaris africana* ENGL., *Ceiba pentandra* GAERTN. f., *Chlorophora excelsa* BENTH. et HOOK. f., *Khaya anthotheca* C. DC., *Terminalia superba* ENGL. et DIELS., *Triplochiton scleroxylon* K. SCHUM. and others.

In the moist deciduous tropical forests the deciduous mountain forests of eastern Africa may also be included. They comprise such economically important hardwoods as *Cephalosphaera usambarensis* WARB., *Entandrophragma deiningeri* HARMS, *Ocotea usambarensis* ENGL. and the coniferous species of *Juniperus procera* HOCHST. ex A. RICH. and species of the *Podocarpus* L'HÉRIT. genus (Fig. 15).

Typical moist deciduous (monsoon) tropical forests extend in southern and south-eastern Asia from India, Nepal, Bhutan and Bangladesh to Burma, Thailand, Laos, Cambodia and Vietnam to Indonesia (northeastern Java and southern Irian). Trees of the upper storey attain a height of 30–40 m. The most important tree species of the moist deciduous forests of India, Nepal and Bangladesh are *Tectona grandis* L. f. and *Shorea robusta* Gaertn. f. (Champion, 1936). In Burma and Thailand the main commercial tree species include *Tectona grandis* L. f. and species of the *Dalbergia* L. f., *Terminalia* L., *Vitex* L. and *Xylia* Benth. genera. In Indochina there are the following important tree species belonging to the Dipterocarpaceae family: *Dipterocarpus alatus* Roxb., *Hopea ferrea* Pierre, *Shorea vulgaris* Pierre, *Vatica bancana* Scheff., and from other families there are, for example, *Lagerstroemia angustifolia* Pierre, *Afzelia xylocarpa* Craib., *Pterocarpus macrocarpus* Kurz and species of the *Diospyros* L. and *Xylia* Benth. genera. Economically important tree species belonging to the moist deciduous tropical forests in Indonesia include, for example, *Eucalyptus alba* Reinw., *Malaleuca leucadendron* L. and the pine *Pinus merkusii* Jungh. et De Vriese.

In Latin America moist deciduous tropical forests occur mainly in Brazil, Bolivia, Venezuela and in some regions of Central America. Commercial tree species in this part of the world include species of the *Andira* Engl., *Dalbergia* L. f. and *Tabebuia* Gomez genera, in addition *Cedrela odorata* L., *Swietenia mahagoni* Jacq., and among coniferous species mainly of *Pinus caribaea* Morelet and *P. oocarpa* Schiede are represented (Beard, 1955).

2.3.3 Dry deciduous tropical forests

Closed dry deciduous tropical forests occur in a variety of vegetation types, mainly on hills and plateaus of the tropics where air temperature in the warmest months varies around 40–46° C. Annual precipitation is usually 600–1,200 mm with a rainy season that lasts from 5 to 9 months.

The dry deciduous tropical forests (also called dry monsoon forests, woodland savanna or open dry forests) consist mainly of hardwoods which shed their foliage during the dry season. The trees grow to a height of 10–25 m and have a diameter of 40–70 cm and they often have boles of bad shape. The undergrowth is discontinuous in its cover and is made up of deciduous and evergreen shrubs, grasses and xerophytic herbs.

In extreme dry and hot climatic conditions where there is an annual precipitation of 300–600 mm, dry deciduous tropical forests turn into open deciduous thorn-bush (Fig. 16) or into succulent growth where trees attain a height of 3–12 m and a diameter of 15–30 cm. In the undergrowth there are grasses, xerophytic shrubs and succulents (Fig. 17).

In Africa dry deciduous tropical forests occur to the north of the equator in the Sudan belt. Characteristic tree species in this region are species of the *Albizia* Durazz., *Parkia* R. Br. and *Uapaca* Baill. genera.

A characteristic tree species of lowland dry tree savannas almost anywhere in Africa to the south of the Sahara is the baobab – *Adansonia digitata* L., a broad-trunked, bottle-shaped tree which forms sporadically coherent stands (Fig. 18).

Fig. 16. Dry deciduous thorny tropical forest. Usangu plane, Tanzania.

Fig. 17. Succulent of the *Caralluma* genus with very dark flowers. Umba-steppe. Tanzania.

Typical trees of dry deciduous tropical forests in Africa are species of the *Acacia* WILLD., *Albizia* DURAZZ., *Combretum* LOEFL., *Commiphora* JACQ., *Diospyros* L. and *Terminalia* L. genera as well as *Tamarindus indica* L. and other tree species

Fig. 18. Stand of *Adansonia digitata*. Valley of the Ruaha river, Tanzania.

(Figs 19, 20). In East Africa economically important tree species, though of low growth, occur on heavy dark soils mostly in lowlands but also on hills, these include blackwood – *Dalbergia melanoxylon* GUILL. et PERR. (Fig. 21) and muhuhu – *Brachylaena hutchinsii* HUTCH.

Fig. 19. Tree savanna with *Acacia tortilis.* Serengeti National Park, Tanzania.

Fig. 20. Woodland savanna with species of *Acacia, Commiphora* and *Lannea* genera. Serengeti National Park, Tanzania.

24

Fig. 21. Stand of *Dalbergia melanoxylon,* height of trees 10–12 m. Region of Kilwa, Tanzania.

Fig. 22. The palm-tree *Borassus aethiopum* forms open stands of 20–25 m height. Busenge Forest Reserve, Tanzania.

Fig. 23. Dry deciduous tropical forest of the miombo type with species mainly of the *Brachystegia* genus. Region of Biharamulo, Tanzania.

The palm-tree *Borassus aethiopum* MART. is locally common in eastern and central Africa. It forms open stands of 20–25 m in height and occurs scattered from sea-level to about 1,200 m altitudes, especially in low lying areas (Fig. 22).

South of the equator the dry deciduous tropical forests are floristically richer. The most important type of this forest type of formation is the miombo forest which extends over wide areas of Angola, Zaire, Tanzania, Mozambique, Zambia and Zimbabwe. The main economic tree species of the miombo forest in Zambia and Zimbabwe is *Baikiea plurijuga* HARMS (Ernst, 1971). In other countries species of the *Brachystegia* BENTH. (Fig. 23) and *Isoberlinia* CRAIB et STAPF genera form the most important communities. A very valuable tree species in the miombo forest is *Pterocarpus angolensis* DC. The trees and shrubs show a flowering maximum during the dry hot season. Fruiting shows a seasonal rhythm with a maximum during hot dry season (Malaisse, 1978).

In India and Burma *Tectona grandis* L. f., *Anogeissus latifolia* WAHL., *Diospyros tomentosa* ROXB. and *Hardwickia pinnata* ROXB. are characteristic tree species of dry deciduous forests. In Thailand dry deciduous tropical forests grow in lowlands and on hills to an altitude of 1000 m. In Laos and Vietnam dry deciduous tropical forests occur on poor stony and lateritic soil at higher altitudes. The most important tree species of these forests are included in the Dipterocarpaceae family (*Dipterocarpus tuberculatus* ROXB., *D. intricatus* DYER., *D. obtusifolius* TEYSM., *Pentacme siamensis* KURZ and *Shorea obtusa* WALL.).

In Central America dry deciduous tropical forests occur where there is an annual precipitation of 500 to 1,000 mm. They grow in Mexico, mainly on the Yucatan Peninsula, in Costa Rica, Honduras and in the western part of the Greater Antilles. Characteristic tree species are *Bursera simaruba* SARG., *Acacia farnessiana* (L.) WILLD., *Swietenia humilis* ZUCC. and species of the *Prosopis* L. genus. Most tropical timber species are not uniformly distributed throughout the Central America including the Caribbean area but may be arranged into several geographical groups (Longwood, 1962).

In South America vast areas of the Brazilian highlands called "cerradão", the southeastern part of Peru and the central part of Bolivia are covered by dry deciduous tropical forests. In Paraguay and in the Chaco region of Argentina there are open dry deciduous tropical forests where trees attain heights of 10–15 m. Among the more important economic tree species of dry deciduous tropical forests in South America are the species knowns as quebracho – which are mainly *Schinopsis lorentzii* ENGL. and *Aspidosperma quebracho-blanco* SCHLECHTD. and species of the *Calophyllum* L., *Tabebuia* GOMEZ, *Prosopis* L. and *Zizyphus* JUSS. genera* (Heinsdijk, 1960).

Dry deciduous tropical forests are important mainly for soil conservation, but also for fuelwood and other locally used products. Dry deciduous tropical forests are

---

* Hereafter Latin names are given, as are commercial names, only for economically important tree species.

heavily damaged by the grazing of cattle and also by almost annually recurring forest fires.

The Central and South America possess about 70% of the world's tropical softwood forests; 32 million ha out of 46 million ha. These forests are mainly situated in Central America and the Caribbean Islands, more than 5 million ha of softwood tropical forests exist in southeastern Brazil (Lanly and Clement, 1979).

3 Importance of tropical forests as a source of timber

3.1 World timber production

According to the Yearbook of Forest Products (FAO, Rome 1986) the current world-wide annual felling and consumption of timber amounts to 3,165 million m³ (1985), i.e. more than 1 m³ per hectare of closed forest. In 1985 this global timber production included 1,502 million m³ of industrial wood (47% of the global production) and 1,663 million m³ of fuelwood (53%).

The total global annual timber production includes 1,243 million m³ (39%) of coniferous wood and 1,922 million m³ (61%) of broad-leaved wood. Whereas 83% of the world production of coniferous timber is processed by industry, only 26% of the world production of broad-leaved wood is so used.

The development of timber production globally during the 20 years preceding 1985 is given in Table 2. Total timber production increased from 2,251 million m³ in 1966 to 3,165 million m³ in 1985 which is up by 914 million m³, i.e. by 40%. In this twenty-year period the production of industrial wood increased by 348 million m³, i.e. by 30%, and the production of fuelwood increased by 566 million m³, i.e. by 52%.

The share of industrial wood in global timber production was 51% in 1966, this increased gradually to 53.5% by 1972. As a consequence of the steep price rises of raw materials (including oil and timber) in 1973 the share of industrial wood in the total timber production declined gradually to 47.5% by 1985. Also, the stagnation of economic growth at the beginning of the eighties led to a decline in the production of industrial wood (Table 2).

If the annual timber production is separated into groups of countries according to their national economy, viz:
– countries with developed economies,
– countries with central planning,
– developing countries,
then the largest share in total timber production goes to the developing countries with 46% (1,460 million m³) followed by countries with developed economies rep-

Table 2. World production of roundwood (Yearbook of Forest Products, FAO, Rome, 1974, 1986)

Year	Industrial wood		Fuelwood		Total roundwood
	$1,000 \, m^3$	%	$1,000 \, m^3$	%	$1,000 \, m^3$
1966	1,154,004	51.3	1,096,735	48.7	2,250,739
1968	1,205,205	52.1	1,107,812	47.9	2,313,017
1970	1,276,016	53.3	1,120,089	46.7	2,396,105
1972	1,313,302	53.5	1,140,204	46.5	2,453,596
1974	1,351,006	51.4	1,277,308	48.6	2,628,314
1976	1,370,864	50.7	1,335,477	49.3	2,706,341
1978	1,423,534	50.6	1,387,792	49.4	2,811,326
1980	1,445,213	49.1	1,496,743	50.9	2,941,956
1982	1,366,029	46.5	1,575,023	53.5	2,941,052
1984	1,494,864	47.8	1,632,265	52.2	3,127,129
1985	1,502,047	47.5	1,662,972	52.5	3,165,019

resenting 30% (958 million m^3) and finally by countries with central planning which account for 24% (747 million m^3) of the global timber production. The group of countries with central planning include the COMECON countries with the exception of Cuba, the People's Republic of China and the People's Democratic Republic of Korea.

The countries with the largest area of closed productive forests are the U.S.S.R. with 791 million ha, Brazil with 396 million ha, Canada with 264 million ha, the U.S.A. with 192 million ha and mainland China with 125 million ha.

The countries with the largest annual timber production in 1985 were the U.S.A. with 448 million m^3, the U.S.S.R. with 356 million m^3, China with 263 million m^3, India with 245 million m^3 and Brazil with 226 million m^3. The largest producers of industrial wood in 1985 were the U.S.A. with 346 million m^3, the U.S.S.R. with 275 million m^3, Canada with 165 million m^3, China with 93 million m^3 and Brazil with 58 million m^3.

The largest consumers of industrial wood in 1985 were the U.S.A. (327 million m^3), U.S.S.R. (258 million m^3), Canada (165 million m^3), China (107 million m^3) and Japan (74 million m^3).

In some countries, mainly in those with developed forest industries, the indigenous timber resources do not meet requirements. Here, fast dwindling stocks of timber require that there are imports from countries which have ample timber supplies. After World War II such imports came more and more from tropical regions (Pringle, 1976, 1979).

General world-wide data on growing timber stocks are incomplete and are often obtained by rough estimates. The same holds true for information published on timber production and its utilization in various countries. Timber extraction for local consumption is not always recorded especially in developing countries.

3.2 Share of tropical forests in world timber production

Of the total of 3,165 million m³ of timber production world-wide in 1985, 1,459 million m³ (i.e. 46%) were extracted from tropical forests in developing countries. This timber volume included 253 million m³ of industrial wood and 1,206 million m³ of fuelwood, or 17 and 83%, respectively, of the total timber production in tropical forests.

From the total production of industrial wood world-wide in 1985 (i.e. 1,502 million m³) 801 million m³ or 53% belongs to the countries with developed economies, 448 million m³ or 30% belongs to the countries having central planning, and 253 million m³ or 17% belongs to the developing countries.

The development of production of industrial wood and fuelwood in the developing countries for a twenty-year period ending in 1985 is given in Table 3.

Table 3. World production of roundwood in developing countries (Yearbook of Forest Products, FAO, Rome, 1974, 1986)

Year	Industrial wood		Fuelwood		Total roundwood
	1,000 m³	%	1,000 m³	%	1,000 m³
1966	126,087	14.5	741,875	85.5	867,962
1968	143,500	15.6	774,723	84.4	918,223
1970	159,639	16.6	801,248	83.4	960,887
1972	173,822	17.4	825,541	82.6	999,363
1974	185,506	16.4	944,944	83.6	1,130,450
1976	207,357	17.2	998,140	82.8	1,205,497
1978	226,158	18.1	1,024,425	81.9	1,250,583
1980	247,318	18.6	1,078,867	81.4	1,326,185
1982	242,299	17.7	1,129,235	82.3	1,371,534
1984	253,396	17.7	1,180,021	82.3	1,433,417
1985	253,316	17.4	1,206,012	82.6	1,459,328

Total timber production in the developing countries during this time increased from 868 million m³ in 1966 to 1,459 million m³ in 1985, i.e. by 591 million m³ or 68%. In this period, the production of industrial wood increased from 126 million m³ in 1966 to 253 million m³ in 1985, i.e. by 127 million m³ or 101%, and the production of fuelwood increased by 464 million m³ or 63%.

Total timber production from tropical forests increased from 39% of world timber production in 1966 to 46% of world timber production in 1985. The share of rough industrial wood extracted from forests in developing countries is low and increased during the twenty-year period from 14 to 17% of world production of industrial wood. As shown by data in Table 2, the price increase for raw materials in the early seventies and the world-wide economic stagnation in the early eighties had an adverse effect on the production of industrial wood in the developing countries. It should be noted that industrial utilization of timber is correlated to high standards of a national economy and to the development of forest industries in the respective country. Government policy in many developing countries at present is to set up domestic forest industries, to process felled timber in the homeland and to restrict the exportation of tropical logs (Sommer, 1976).

The permanently increasing human population in developing countries causes a constant increase in the consumption of fuelwood. At the time of a universal energy crisis in the world, wood as a source of thermal energy acquires an outstanding importance in those developing countries which lack their own oil reserves or other energy.

Considering the rate of tropical forest depletion during the past years and the scope of various national development plans, it is estimated that about 7.5 million ha of closed forests and 3.8 million ha of open forests will disappear annually, which comes to an overall rate of 0.6% deforested per year. Every year 4.4 million ha of undisturbed closed forest are subjected to timber exploitation. Therefore, to the 11.3 million ha of tropical forest lost annually through deforestation caused by population growth and subsistence agriculture a further 4.4 million ha of closed forest affected by new loggings must be added for a total of 15.7 million ha annually. The total annual rate of tropical forest reduction and disturbance is 0.8% (Steinlin, 1982).

The volume of fuelwood obtained in developing countries in 1985 attained 1,206 million m^3 or 73% of the world production of fuelwood which equalled 1,663 million m^3. In 1966 the share of fuelwood production in developing countries was 68% (742 million m^3) of world fuelwood production (1,097 million m^3). The area of tropical forests declines as a consequence of excessive fellings of both industrial wood and fuelwood, but mainly it declines because of the pressure of local human populations for forest land to be used for the cultivation of agricultural crops. Generally the people use the land around the forest in two basic ways; raising crops mostly for food and the use of grazing land for cattle and other domestic animals. Agri-silviculture is a production technique which combines the growing of agricultural crops with simultaneously raised and protected forest stands (Chaudhry and Silim, 1980).

The total production of 1,459 million m^3 of tropical timber in the developing countries in 1985 was distributed among continents as follows:

Africa	418 million m^3 or 29%,
Asia and the Pacific	681 million m^3 or 47%,
Latin America	360 million m^3 or 24%.

In Africa, countries with the largest timber production in 1985 included Nigeria (95 million m^3), Tanzania (45 million m^3) and Ethiopia (38 million m^3). Most industrial wood in Africa in 1985 was produced by Nigeria (8.0 million m^3), Ivory Coast (4.5 million m^3) and Cameroon (2.8 million m^3). Although Zaire has the largest area of closed tropical forests in Africa, namely about 106 million hectares, total timber production in this country in 1985 was 30.5 million m^3 including only 2.5 million m^3 of industrial wood.

In Asia and the Pacific the largest producers of tropical timber in 1985 were as follows: India (245 million m^3), Indonesia (149 million m^3), Thailand (41 million m^3) and Malaysia (40 million m^3). The largest amounts of industrial wood in the same year were produced by Malaysia (32.1 million m^3), Indonesia (26.8 million m^3) and India (22.6 million m^3).

The average volumes extracted per hectare in the undisturbed productive closed broad-leaved forests are relatively higher in tropical Asia (31 m^3 per hectare) than in tropical America (8.5 m^3 per hectare) and tropical Africa (13.5 m^3 per hectare). This is because of richness in commercial species of the Dipterocarp forests of Southeast Asia (Lanly, 1982).

In the Pacific, with the exception of Australia and New Zealand, the surface of closed moist tropical forests is estimated to be 40 million ha. The largest timber resources are found in New Guinea and Papua where annual production of industrial tropical wood amounts to about 2 million m^3 (1985).

The total area covered by coniferous forests is about 8.4 million ha, with about two-thirds of these considered productive. They are essentially confined to the Himalayan belt in southern Asia (Rao and Chandrasekharan, 1983).

In Latin America, Brazil with 226 million m^3, Mexico with 21 million m^3 and Colombia with 17 million m^3 are the largest producers of tropical timber. Brazil, the world's second largest producer of tropical timber after India, also has the largest area of closed tropical forest in the world (396 million ha) and, in the basin of the Amazon river, the largest expanse of evergreen moist tropical forest which covers an area of about 300 million ha. It should be noted that in Brazil about 1.5 million ha of productive forest is deforested annually. Brazil has the world's largest area of plantations of tropical and subtropical tree species amounting to 3.8 million ha (1980), and about 350 thousand ha of land is reafforested annually. In Brazil the steady decline of the natural tropical forest and the ensuing gradual deterioration of forest soils are continuing.

In Latin America the surface of coniferous tropical and subtropical forests amounts to about 25 million ha including over 15 million ha of productive forests. The largest surface of coniferous productive forests – 11.7 million ha – occurs in Mexico.

The production of industrial wood in Latin America was 95 million m^3 in 1985. Brazil is the largest producer of industrial wood in Latin America with 58 million m^3 representing 61% of the total production of industrial wood in Latin America. This is followed by Chile with 9.4 million m^3 and Mexico with 7.5 million m^3 of industrial wood. In all three of these countries a large share goes to coniferous tropical industrial wood which represents 40% in Brazil (23.4 million m^3), 92% in Chile (8.3 million m^3) and 89% in Mexico (6.7 million m^3). In the observed twenty-year period, production of industrial wood increased in Brazil from 18 million m^3 (1966) to 58 million m^3 (1985), it is thus up by 40 million m^3 or 222%. In Chile it increased from 4.1 million m³ to 9.4 million m³, thus up by 5.3 million m³ or 129%, and in Mexico it increased from 3.7 million m³ to 7.5 million m³, thus up by 3.8 million m³ or 103%.

From the point of view of production of tropical industrial wood, evergreen moist and semi-deciduous moist tropical forests with their rich occurrence of various commercial and other less well-known tree species, are of the greatest importance for all tropical regions of the world. Moist tropical evergreen and semi-deciduous tropical forests cover about 900 million ha, and the production of tropical sawlogs and veneer logs, at present about 140 million m³, represent about 60% of world production of hardwood logs (Fontaine, 1986).

In African tropical forests, about 220 widely known commercial tree species are utilized. They include about one hundred species which occur more or less regularly on world timber markets.

In tropical Asia timber from about 300 different tree species is sold in the market, many of them under the same commercial designation. Thus, for example, 92 species of *Shorea* and 3 species of *Parashorea* are being designated commercially as "meranti". More than 200 tree species included in 11 genera are embraced in 10 commercial names which have the largest annual production in southeastern Asia.

In Latin America about 470 different tree species from tropical forests have been recorded on the timber market including 210 species, which occur in reasonable quantity.

For better information on the production of tropical logs for the 10 most frequently extracted commercial tree species data are given in Tables 4–6 according to the main tropical regions of the world in 1973 which was the year of the highest boom in the world timber market.

From the commercial point of view, all tropical tree species with an annual production exceeding 1,000 m³ are considered to be commercial species. The remaining species which have an annual production of less than 1,000 m³ are properly called species of minor use.

In the period from 1951 to 1973 in Africa, for example, 105 commercial tree species and 112 species of minor use were recorded. From among them, okoumé *(Aucoumea klaineana)* and wawa *(Triplochiton scleroxylon)* represented 31% of tropical log production in Africa.

Other widely used commercial tree species in Africa include the African mahoganies (tree species of the *Entandrophragma* and *Khaya* genera), iroko *(Chlorophora excelsa)* and limba *(Terminalia superba)*.

In tropical Asia the most widely extracted tree species include species of the *Shorea, Parashorea, Pentacme* and *Dipterocarpus* genera; they are known under the commercial designation of meranti, lauan, seraya and yang.

In Latin America species of the *Virola, Swietenia, Cedrela* and *Phoebe* genera belong to the most frequently extracted commercial tree species.

Forest products other than wood (minor forest products) can be classified into food products (fruit, oil producing seeds, honey, mushrooms), fibres (bamboo, rat-

Table 4. Production of tropical logs of the 10 most utilized tree species in western and equatorial Africa in 1973 (data given in 1,000 m³) (Erfurth and Rusche, 1976)

Designation of tree species		Logs production in various countries								Total
Commercial designation	Latin name	LIB	I.C.	GHA	NIG	CAM	GAB	CGO	ZAI	
obeche	*Triplochiton scleroxylon*	21	1,109	520	229	69	–	5	7	1,960
okoumé	*Aucoumea klaineana*	–	–	–	–	–	1,571	211	–	1,782
utile (sipo)	*Entandrophragma utile*	49	430	107	2	26	9	7	31	661
African mahogany (acajou)	*Khaya* spp. (3 species)	3	195	161	136	18	19	28	26	586
sapele (sapelli)	*E. cylindricum*	7	172	190	7	84	–	61	31	552
limba	*Terminalia superba*	–	51	1	265	16	39	65	24	462
iroko	*Chlorophora excelsa*	9	184	95	17	18	–	6	20	349
tiama	*E. angolense*	2	179	83	6	4	5	3	26	308
ekki, red ironwood (azobé)	*Lophira alata*	16	5	1	1	261	–	–	–	284
makoré	*Dumoria* spp. (2 species)	20	166	56	–	2	15	15	–	274

Countries: Liberia, Ivory Coast, Ghana, Nigeria, Cameroon, Gabon, Congo, Zaire.

Table 5. Production of tropical logs of the 10 most utilized tree species in southeastern Asia in 1973 (data given in 1,000 m³) (Erfurth and Rusche, 1976)

Designation of tree species		Logs production in the various countries					Total
Commercial designation	Latin name	MAL	SAR	SAB	IND	PHI	
meranti	*Shorea* (92), *Parashorea* (3)	3,312	455	–	13,901	–	17,688
Philippine mahogany	*Shorea* (5), *Parashorea* (1), *Pentacme* (1)	–	–	–	–	7,084	7,084
red seraya	*Shorea* (32)	–	–	5,934	–	–	5,934
keruing	*Dipterocarpus* (52)	2,042	85	1,205	1,417	–	4,749
kapur	*Dryobalanops* spp. (9)	1,313	85	1,867	251	–	3,516
white seraya	*Parashorea* spp. (5)	–	–	3,091	–	–	3,091
ramin	*Gonystylus* spp. (8)	–	640	–	2,107	–	2,748
apitong (yang)	*Dipterocarpus* spp. (10)	–	–	1	–	1,597	1,597
alan	*Shorea albida*	–	1,399	–	–	–	1,399
yellow seraya	*Shorea* spp. (10)	–	–	949	–	–	949

Countries: Malaya, Sarawak, Sabah, Indonesia, Philippines.

Table 6. Production of tropical logs of the 10 most utilized tree species in South America in 1973 (data given in 1,000 m³) (Erfurth and Rusche, 1976)

Designation of tree species		Logs production in various countries							Total
Commercial designation	Latin name	BRA	BOL	PER	EQU	COL	VEN	GUI	
virola	*Virola* spp. (7)	1,105	3	1	5	510	–	38	1,662
mahogany (caoba)	*Swietenia* spp.	204	150	34	–	–	23	–	411
cedro	*Cedrela* spp. (2)	107	1	130	13	12	–	2	265
balsa	*Ochroma lagopus*	–	–	–	246	–	–	–	246
andiroba	*Carapa* spp. (2)	151	–	–	13	34	–	21	219
orey (sajo)	*Campnosperma panamensis*	–	–	–	–	182	–	–	182
louro inhamuy	*Ocotea cymbarum*	149	–	–	–	–	–	–	149
mahot coton (saqui-saqui)	*Bombacopsis quinatum*	–	–	–	–	1	124	–	125
greenheart	*Ocotea rodiaei*	–	–	–	–	–	–	117	117
espavel (mijao)	*Anacardium excelsum*	–	–	–	6	8	92	–	106

Countries: Brasil, Bolivia, Peru, Equador, Colombia, Venezuela, Guiana.

tan), rubber, gums, resins, waxes, tannins and pharmaceutical and cosmetic products. The products of forest animal also are important for local people.

3.3 Share of tropical forests in world timber export

Information on the share taken by developing countries in world timber exports is contained in Table 7 which is derived from the Yearbook of Forest Products (FAO, 1986).

Table 7. Share of developing countries in world export of forest products in 1985 (Yearbook of Forest Products, FAO, Rome, 1986)

Forest products	World	Developing countries	
	1,000 m³		%
Coniferous logs	32,685	1,429	4.4
Broad-leaved logs	30,095	26,830	89.2
Coniferous sawn wood	73,575	1,480	2.0
Broad-leaved sawn wood	11,717	7,917	67.6
Veneer sheets	1,821	963	52.9
Plywood	8,480	5,920	69.8
Particle board	2,276	250	11.0
	1,000 t		%
Wood pulp	21,772	1,731	8.0
Paper and paper-board	40,292	1,384	3.4

Coniferous timber is mainly produced in countries of the temperate zone, therefore the share of tropical and subtropical developing countries in the world export of coniferous sawn timber and logs is small (2–4%). Chile is the largest exporter of tropical coniferous logs and sawn timber with 94% (1.3 million m³) and 50% (0.7 million m³), respectively, of the world export. Brazil, formerly the largest exporter of tropical coniferous sawn timber, restricted the export of this commodity from 1.3 million m³ in 1968 to 72,000 m³ in 1985.

The share of tropical developing countries in the world export of broad-leaved logs is very important, being 89% in 1985. Developing countries also have an important share in the export of broad-leaved sawn timber which represented 68% of the world export in 1985.

Malaysia is the largest exporter of tropical broad-leaved logs in the world. This country exported 19.8 million m³ of logs, i.e. 70% of the world export of tropical broad-leaved logs in 1985. Indonesia, recently the second largest exporter of tropical logs in southeastern Asia, substantially reduced the export of broad-leaved tropical logs, from 19 million m^3 in 1978 to about 1 million m^3 in 1985. By expanding her forest industries, Indonesia turned to the export of plywood and tropical sawn timber (3.6 million m^3 and 2.2 million m^3, respectively, in 1985).

In Africa the largest exporters of tropical logs at present include the Ivory Coast with 1.3 million m^3, Gabon with 1.1 million m^3 and Cameroon with 750,000 m^3. Zaire, the country with the largest reserves of tropical timber in Africa, exported only 100,000 m^3 of logs and 26,000 m^3 of sawn timber in 1985.

Although the largest resources of tropical timber occur in South America, the developing countries of Latin America have a share of only 5% in the world commerce of tropical timber and products derived from it. The export of broad-leaved logs is negligible and was about 60,000 m^3 in 1985. The volume of exported tropical broad-leaved sawn timber from Latin American countries amounted to 900,000 m^3 or about 8% of the world export.

The largest exporter of tropical broad-leaved sawn timber in Africa is the Ivory Coast with 414 thousand m^3 (1985). In Asia the largest exporters are Malaysia with 2,740 thousand m^3 and Indonesia with 2,166 thousand m^3. In Latin America Brazil with 420 thousand m^3 and Paraguay with 350 thousand m^3 of sawn timber are the biggest exporters.

Table 7 contains data on both broad-leaved and coniferous veneer and plywood. The high share of developing countries in the world export of veneer (53%) and plywood (70%) is as a consequence of a developed plywood industry in certain developing countries, such as in Brazil, Mexico, Nigeria and especially in parts of southeastern Asia. The largest exporters of veneer in 1985 were: in Africa the Ivory Coast with 65 thousand m^3 and Congo with 50 thousand m^3; in Latin America Paraguay with 64 thousand m^3 and Brazil with 52 thousand m^3; and in Asia Malaysia with 415 thousand m^3 and Indonesia with 108 thousand m^3. The largest exporters of plywood in 1985 were in Asia: Indonesia with 3,575 thousand m^3, Singapore with 629 thousand m^3, Malaysia with 363 thousand m^3 and the Philippines with 269 thousand m^3; in Africa Gabon with 46 thousand m^3; and in Latin America Brazil with 230 thousand m^3.

The share of developing countries in the world export of particle boards and pulpwood is negligible, representing about 1–2%. The larger share of fibreboard in world export (11%) is influenced by the export from South American countries such as Brazil (171 thousand m^3), Chile (32 thousand m^3) and Argentina (22 thousand m^3) in 1985.

The main exporters of cellulose in 1985 were Brazil which exported 930 thousand tons, and Chile which exported 503 thousand tons. This represented 85% of all cellulose exported from developing countries. In the world export of paper and cardboard an important role was played in 1985 by South American countries such as Brazil (543 thousand tons) and Chile (134 thousand tons) and also by countries in southeastern Asia including South Korea (109 thousand tons), Singapore (83 thousand tons), and Hong Kong (71 thousand tons). These five developing countries provide 83% of the export of paper and cardboard from all developing countries in the world, but this amounts to only 3% of the world export of these commodities.

A comparison of statistical information published by United Nations Conference on Trade and Development (UNCTAD) and Food and Agriculture Organization of the United Nations (FAO) gives some idea of the part played by timber and timber products between developing countries. Between 1970 and 1983 it rose from 11 to 35 million m^3 of roundwood equivalent. The trade in wood and wood products between developing countries has expanded faster than exports from developing countries to developed countries. Asian growth in this area seems to be stronger than that of South American countries, for which horizontal trade represents between one-third and one-half of their exports (Buttoud and Hamadou, 1986).

The most frequent tree species exported from Latin America in the form of logs and sawn timber include species of the *Cedrela* genus (ceder, cedrela, cedro – about 10 species), of the *Virola* genus (baboen, virola – over 30 species) and of the *Peltogyne* genus (amarants, violetwood, purpleheart – about 10 species). The species of *Swietenia mahagoni* and *S. macrophylla* (caoba, aguano), *Phoebe porosa* (imbuia) and *Ferreirea spectabilis* (sucupira), are also prominent.

From Malaysia and Indonesia logs of the following tree species are mostly exported: *Shorea, Parashorea* and *Pentacme* (meranti, lauan, seraya), species of the *Dryobalanops* genus (kapur), and species of the *Dipterocarpus* genus (keruing, yang). And in the form of sawn timber the following tree species are mostly exported: species of the *Gonystylus* genus (ramin), species of the *Shorea* and *Parashorea* genera (meranti, lauan), and species of the *Dipterocarpus* genus (keruing).

Africa is the main supplier of tropical timber to European countries. Until recently timber of very large dimensions with excellent technological properties has been imported from Africa. Until 1973 tropical timber supplies from Africa accounted for 90% of European consumption. From 1951 to 1973 the number of tree species exported from Africa to Europe (in a volume exceeding 10,000 m^3 annually) increased from 18 to 35 and at present it is still increasing.

Up to now the two most frequently exported tree species, namely *Triplochiton scleroxylon* (wawa or obeche) and *Aucoumea klaineana* (okoumé), represented 39% of the total export of logs and sawn timber from Africa in 1973 (Bertrand, 1974).

After the steep price increase of tropical timber in 1973 (by 90% on average in Africa) the export of tropical timber, mainly logs, decreased in the following years and gradually turned from very valuable commercial tree species to less well-known and relatively cheaper tree species. As an illustration, the list of the 20 most widely exported tree species in log form from the Ivory Coast in 1971 and 1981 is given in Table 8. These species represented 95% of the total log export in 1971 and 90% in 1981.

According to data published by the European Economic Commission in Geneva (ECE, Timber Bulletin for Europe, Geneva, 1978) the largest importers of tropical

Table 8. The twenty most widely exported tree species in log form from the Ivory Coast in 1971 and 1981 (Bois et Forêts des Tropiques, 1972; Tropical Forest Products in World Timber Trade, Monthly Bulletin, FAO, 1982)

Tree species	Export of logs in 1,000 m³	
	in 1971	in 1981
Triplochiton scleroxylon	933	375
Entandrophragma utile	564	56
Khaya ivorensis	172	81
Dumoria heckelii	161	35
Entandrophragma angolense	147	77
Entandrophragma cylindricum	146	47
Mansonia altissima	105	40
Pycnanthus angolensis	90	89
Lovoa trichilioides	81	20
Chrysophyllum africanum	79	232
Entandrophragma candollei	67	26
Chlorophora excelsa	63	116
Terminalia ivorensis	60	71
Tarrietia utilis	57	125
Pterygota macrocarpa	37	67
Antiaris africana	36	151
Ceiba pentandra	–	144
Nesogordonia papaverifera	–	130
Terminalia superba	–	114
Mitragyna ciliata	–	103
Total	2,798	2,099
Total log export of all species	2,933	2,343

timber in the world in 1973 included Japan with 28.9 million m³, the U.S.A. with 7.1 million m³ and the countries of western Europe with 15.7 million m³.

As a consequence of the price increase of tropical timber in 1973 the export of tropical timber, especially of logs, decreased in the following years. In 1976 Japan imported a total of 22.7 million m³, the U.S.A. 5.7 million m³ and western Europe 13.8 million m³. Among the largest importers of tropical timber in 1976 were also South Korea with 5.5 million m³ and Taiwan with 4.6 million m³ (ECE, 1981; UNCTAD, 1982).

Also, today, Japan is maintaining its overriding position in the world as to the import of tropical timber. In 1985 it imported, mainly from the regions of southeast Asia and the Pacific, 13.3 million m³ of tropical logs, 920 thousand m³ of sawn timber, 70 thousand m³ of veneer sheets, 320 thousand m³ of plywood and 330 thousand m³ of pulpwood and chips of tropical wood. The largest importers of tropical logs into Europe in 1985 were France with 1,050 thousand m³ and Italy with 650 thousand m³. The largest amount of sawn timber was imported from the developing countries into Europe by F. R. G. with 450 thousand m³ and by Great Britain with 430 thousand m³. Most plywood was imported by Great Britain (460 thousand m³) and the Netherlands (170 thousand m³). The U.S.A. reduced radically the import of tropical logs in 1985 to only 30 thousand m³, but the imports of veneer sheets rose to 112 thousand m³, plywood to 1,300 thousand m³ and tropical sawn timber to 480 thousand m³.

As to the price of tropical timber, it should be noted that it had been stable throughout a long period of years. The price formation depended on supply and demand, and in this process a certain equilibrium has been developed between the increasing consumption of tropical timber in European countries and the rising production of tropical timber in certain African countries. The price increase of tropical timber was justified regardless of the general price increase of raw materials, especially that of oil in 1973 (ECE, 1974). Timber logging receded farther away from ports, and its progressing towards the interior caused an increase in transportation and operational costs, which was combined with increases in wages and various licence and export charges.

Also of no small influence on the increasing price trends of tropical logs are the restriction or prohibition of the export of tropical logs in many developing countries, and the obligation of logging concessionaires to supply logs to the developing local woodworking industry. Such is the situation in most Latin American countries including Brazil, in southeastern Asia this includes Malaya, Thailand, the Philippines, and recently Indonesia; and in Africa, it includes Nigeria, Ghana, and Zaire. The Ivory Coast which is the largest exporter of tropical logs in Africa, restricts likewise the export of logs; the reason behind this policy being the fact that this country has lost about 70% of its commercial forests in the last 25 years by ruthless felling.

Czechoslovakia has been importing tropical logs since 1920. In the pre-war period the maximum annual import of tropical timber attained 16 to 18 thousand m³. The development of imports of tropical logs into Czechoslovakia from 1960 to 1985 is shown in Fig. 24 (according to unpublished information of the Foreign-Trade Organization Ligna, Prague). After the World War II, and mainly since 1960, the imports of tropical logs have gradually increased and attained 35 thousand m³ in 1985.

In Czechoslovakia tropical timber is utilized mainly for the production of decorative veneer and plywood, for furniture, television boxes, radio receivers, panelling, musical instruments, sports equipment and for other purposes. In economically developed countries the share of utilization of tropical timber in construction is about 45%, in furniture 40%, in transport 10% and for other purposes about 5%.

Czechoslovakia has been traditionally importing tropical logs from countries in equatorial and western Africa. This follows from the geographical situation of Czechoslovakia, but also from the traditional orientation towards certain European ports (Hamburg, Bremen, Rotterdam). This orientation has to a considerable degree also influenced the species composition of imported tropical logs. According to Pracna (1978) over 60 tropical tree species representing a volume of 338 thousand m³ of tropical logs were imported into Czechoslovakia from 1959 to 1978.

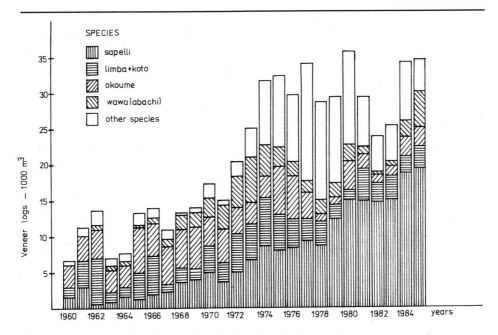

Fig. 24. Development of the import of tropical logs into Czechoslovakia.

The Czechoslovak import of tropical timber includes four main tree species which until recently pertained to the 10 most frequently exported African commercial tree species; they are:

Entandrophragma cylindricum	– sapele,
Aucoumea klaineana	– okoumé,
Terminalia superba	– limba,
Triplochiton scleroxylon	– wawa (obeche).

During the previously mentioned period of 20 years these four tree species represented 79% of the total import of tropical logs into Czechoslovakia. A high percentage of these imports still belongs to one of the most valuable commercial tree species, namely to sapele, the import of which into Czechoslovakia represented 14,691 m³ in 1980 and 16,342 m³ in 1981. This represents more than 50% of the total import of tropical logs.

Although the import of the less well-known commercial species into Czechoslovakia increased (mainly due to the species koto – *Pterygota macrocarpa* which amounted to 6,160 m³ of logs in 1980, and which substituted limba), cheaper tree species like kotibé and red lauan (which may easily replace the commercial species sapele because of their qualities), are rarely used by the woodworking industry of Czechoslovakia.

The present situation requires an adaptation to world development. The technical assistance to developing countries allows for the utilization of their timber resources by establishing common logging and woodworking enterprises. Thus, conditions can

Fig. 25. Terraced fields on originally forested slopes. Region of Mandi, Himachal Pradesh, India.

be created for meeting the demands for tropical timber by capital investment into the logging, transport, processing and export of tropical timber from certain developing countries. An economic plan is required for the yield of raw material from the tropical forest. This is especially true for the lesser-known species which have both less favourable dimensions and differing technological properties when compared with the traditional commercial tree species.

The problem of renewing the national forest resources is of growing concern to many tropical countries (Fig. 25). Reforestation by the concessionaires themselves has in many cases not proved to be successful. Especially smaller and medium-sized operators have neither the necessary facilities and qualified staff nor any sufficiently long-term interest in the future crop that would induce them to engage actively in reforestation. It has been proposed to create an independent reforestation agency which would carry out large-scale plantation programmes on behalf of the concessionaires against payment of a reforestation fee (Schmithüsen, 1976).

At the present stage of forestry development in most developing countries, forest management should generally be understood as a concept of forest resources planning in a particular area. Management or working plans are concerned in the determination of a maximum annual exploitable volume for the main commercial species, infrastructure planning and silvicultural prescriptions.

4 Structure of natural tropical forests

Trees growing in natural tropical forests differ by species, age, growth and dimensions.

Certain tree species have a fast life cycle – 40–100 years, others may live even 1,000 years or more, such as *Balanocarpus heimii* in Malaysia and *Bertholletia excelsa* in Brazil (Hadley and Lanly, 1983). Various tree species have differing ecological demands and this causes the great variation which is to be seen in tree diameter and height, as well as in the dimensions and shape of the crown.

The actual definition of the "structure" of the tropical forest is not universal. Thus Richards (1952) defines the structure of the tropical forest to be the distribution of tree species as biological types (trees, shrubs, herbs, lianas, epiphytes, saprophytes, parasites) and as strata (layers). Other authors, too, understand the forest structure to be the horizontal and vertical architectural order of the forest (Lamprecht, 1954, 1972).

Also Keay (1959), Pierlot (1968) and Rollet (1968) (cited in Rollet, 1974) treat the forest structure in the sense of strata.

Turnbull (1963) discusses the structure of mixed forest stands in terms of the distribution of tree numbers over given area units.

Brüning (1970) uses the term structure in the sense of canopy closure. This is in order to understand the unevenness of tree distribution and the requirements for transpiration of trees in various forest types of Sarawak.

Priesol in Vyskot et al. (1971) understands the term structure to mean the sum total of all quantitative characteristic distinguishing the whole order of the stand. Thus the structure of the stand is given by its origin, its composition and mixture, its age distribution, diameter and height break-up, canopy closure and spatial constitution.

Assmann (1961) defines the term "structure of the selection forest" in the temperate zone (Europe) as the stratification of tree crowns. Each crown layer is characterized by the number of trees, their distribution by diameter breast height and their timber stock for a given area unit.

From a practical point of view, the most important approach seems to be to define the diameter structure of trees in the tropical forest, i.e. the distribution of the frequency of trees over a given area unit by their diameter breast height – D.B.H., or by their diameter above buttresses in the case of certain tree species as is the case mainly in lowland evergreen tropical forests.

One of the first authors to study problems of the diameter structure of trees in the selection forest of the temperate zone was De Liocourt (1898) (cited in Doležal et al., 1969; Rollet, 1974). In virgin fir stands he expressed the decline in the number of trees with an increasing diameter, mathematically by the declining geometrical progression:

$$N_n = A \, q^{-(n-1)},$$

where N_n is the number of trees in the respective diameter class, A the (maximum) number of trees in the first diameter class, q the quotient of the geometrical progression.

Following Liocourt, problems of the diameter distribution of trees were gradually resolved by D'Alverny (1904), Huffel (1926), Schaeffer (1931), Meyer (1933) and other authors.

The declining tendency of tree distribution by their diameter in natural tropical forests in various parts of the world was studied by a number of authors. Le Cacheux (1955) recommended that the diameter distribution of tree frequencies in the tropical forests of Cameroon be expressed by the right side of the normal curve of Laplace–Gauss. Lötsch (Lötsch et al., 1967) expressed the declining tendency of tree distribution by diameter in the deciduous tropical forests of Thailand by an exponential function. Pierlot (1968) does not consider the exponential function as suitable for small diameter classes; he expressed the diameter distribution of trees by a hyperbola function.

The author of this publication firstly uses for the expression of the curve of diameter frequencies, the function:

$$y = k \, e^{-\alpha x},$$

where y is the number of trees in the diameter classes, k, α are constants characterizing the form of the curve, e is the base of natural logarithms, x the diameter classes.

This function was used by Meyer (1933) for the expression of the curve of diameter frequencies in the selection forest of the temperate zone. Although Meyer's function has also been used by certain authors for the expression of the diameter distribution of trees in tropical natural forests, the so-called *beta function* is much better suited to this purpose as has been proved by testing results such as χ^2 – the chi-square test. This function was described for the first time by the mathematician Euler in his paper "Institutionales Calculi Differentiales", St. Petersburg, 1768.

44

The beta function has been recommended for use in expressing the diameter distribution of trees in natural tropical forests by Zöhrer (1969) who also prepared a programme of this function in the computer language FORTRAN IV.

The beta function can be defined by the following integral:

$$B_{(\alpha,\ \beta)} = \text{const} \int_a^b (x-a)^\alpha\ (b-x)^\beta\ \mathrm{d}x,$$

where x is the (variable) argument, i.e. diameter, diameter class, a the lower limit of diameter, b the upper limit of diameter, α, β are first and second exponents of the beta function, const is the multiplication constant, $B_{(\alpha,\ \beta)}$ the total of trees – area below the division curve between a and b.

From the production and commercial point of view information on the tree species composition and diameter structure of trees in tropical forests is of extreme importance in the planning for their rational utilization. The author has obtained information on the tree species composition and diameter structure of trees in natural tropical forests by analysing inventory data on selected forest localities in Ghana, Congo, Gabon and Laos.

The author used the beta function for the mathematical expression of distribution curves of tree diameters for the main biological tree species, biological species grouped together under a commercial species name and lesser-known species. The evaluated inventory data indicated in most cases a declining tendency of the diameter distribution of trees both for identified species and for commercial categories of species in the respective forest formations. The analysed data provide valuable information on the composition, structure and production of the tropical forest in the regions investigated.

Later on basic information will be given about those developing countries where the author has worked. Data on timber resources in these countries are evaluated, and the production possibilities for their tropical forests are discussed. The different diameter structure of trees and the heterogeneity of composition of both economically important and lesser-known species is demonstrated with inventory data from different regions with natural high tropical forests. Mathematically compensated figures of the diameter distribution of trees calculated according to the beta function will be given for various commercial species, for whole areas of natural tropical forests and for relevant forest formations.

4.1 Republic of Ghana

4.1.1 Basic data

The Republic of Ghana lies in western Africa near to the Gulf of Guinea, between 4°45′ and 11°11′ N latitude and 3°15′ W and 1°15′ E longitude. It is surrounded by three French speaking countries: the Ivory Coast, Burkina Faso and Togo. The south borders the Atlantic Ocean for a length of 550 km. With an area of 238,540 km^2 Ghana is almost twice as large as Czechoslovakia. Ghana has about 13 million inhabitants (1984).

The largest city and the country's capital is Accra with 750 thousand inhabitants. The monetary unit is the ₵edi, 1 = 100 pesawas. English is the official language.

Ghana is an agrarian country with a developing industry. About 22% of the country's surface is agricultural land, including shifting cultivation on forest land.

About 65% of the inhabitants depend on agriculture. The main crops are manioc (cassava), yams, bananas, millet, maize and peanuts. Among industrial crops cocoa is the most important. Ghana is the world's largest cocoa producer, yielding about 400,000 tons of cocoa beans annually. Cocoa trees are cultivated in the region of moist tropical forests and will produce a crop for 40–50 years. In addition coconut trees and oil palms are important, and there are hevea (rubber) plantations covering and area of about 10 thousand hectares. Approximately 20% of the working population make a living by fishing, mainly in the coastal region and near Lake Volta.

Mining is the chief industry, gold being the most important mineral. Also of outstanding importance is the mining of diamonds, bauxite and manganese ore. The hydroelectric plant in Akasombo built on the dam of the vast artificial Lake Volta (9,000 km^2) is the main source of energy. Of the total exports from Ghana cocoa, with up to 70% of the export volume, has the greatest share. This is followed by timber and timber products, bauxite and aluminium, gold and diamonds.

About half of the country's land surface is formed by lowlands which rise to an altitude of 150 m, they change to the north into a plateau which lies 200–600 m above sea level. The coastal lowland belt rises to the east to form the Akwapim-Togo Mountains with the highest peaks Jebobo and Torogbani – reaching to 879 and 873 m, respectively, above sea level. This mountain ridge which consists mostly of volcanic rocks continues into Togo under the name of the Atakora Mts.

The important watershed between the Volta river basin and the watercourses flowing to the south is formed by the Kwahu plateau which stretches in a northwesterly direction from the city of Koforidua to the border with the Ivory Coast. The plateau with an altitude of 500–700 m intercepts the rain bearing monsoon winds which blow from the southwest towards the northern regions of Ghana. To the north of the Kwahu plateau the land is undulating with an altitude of 100–400 m where savanna woodland occurs.

Ghana's most important river, the Volta, drains 67% of the country's territory.

Afram and Oti are the chief affluents. Among the more important rivers flowing into the Atlantic Ocean are the Pra, the Ankobra and the Tano. To the southeast of Kumasi is the crater lake of Bosumtwi which is 450 m deep.

Ghana has a tropical climate which is moist and equatorial in the southern part of the country with an annual precipitation of 1,200 to 2,200 mm and no great annual or daily fluctuation of temperature. The average annual temperature in Ghana is 25–29°C. In the northern part of the country the climate is tropical continental with an annual precipitation of 300–1,200 mm and greater ranges of temperature. The Harmattan wind blowing from the Sahara influences the dry and hot climate.

There are great differences in the annual precipitation on Ghana's coast, too. To the east in Accra the annual rainfall is about 720 mm, in the western part of the country on the Atlantic coast annual precipitation is around 2,200 mm. The rainy season lasts from May to September. The coolest month is August, the warmest one is March.

4.1.2 Forests and vegetation formations

Ghana's territory may be divided into two extensive original vegetation formations:
1. the original region of closed high forests (Fig. 26) with an area of 8,226,000 ha, i.e. 34% of the total surface of the country,
2. the original region of savanna woodland with an area of 15,628,000 ha or 66% of the country's surface.

Closed high tropical forests occur in Ghana from the Atlantic coast to 8° N, and in the Akwapim – Togo Mountains to 8°30′ N. This is a region with a hot, moist climate having an annual precipitation of 800–2,200 mm. A characteristic of these formations of moist tropical forests is the uneven age of trees, the great number of various tree species, the differing diameter and height of trees and their varied spatial order.

The region of closed high tropical forests in Ghana includes the following main formations (Taylor, 1962):
– the evergreen tropical forest in the *Cynometra–Lophira–Tarrietia* association,
– the transition belt of the evergreen tropical forest in the *Lophira–Triplochiton* association,
– the semi-deciduous moist tropical forest in the *Celtis–Triplochiton* association (Fig. 27),
– the semi-deciduous moist tropical forest in the *Antiaris–Chlorophora* association.

The evergreen moist tropical forest occurs in the coastal belt in the southwestern part of Ghana where annual precipitation and relative air moisture are the highest. From the ecological point of view this is the most luxuriant and complicated forest formation in the country. It is characterized by the occurrence of a large number of different evergreen tree species. Taylor (1962) mentions 90 mature trees of various species in an area of 48 ha, i.e. approximately two exploitable trees per hectare.

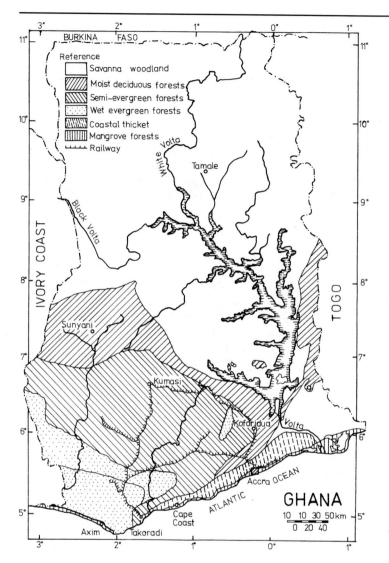

Fig. 26. Main vegetation formations of Ghana.

The evergreen tropical forest in Ghana is characterized by the *Cynometra–Lophira–Tarrietia* association. The highest trees in the uppermost storey attain a height of 40–60 m. Their boles are usually straight, smooth and cylindrical. Some trees have widely spread crowns. Characteristic tree species forming the upper storey include *Cynometra ananta, Daniellia thurifera, Dialium aubrevillei, Erythrophleum ivorense, Parkia bicolor,* and from among commercial species, according to volume representation, *Lophira alata, Piptadeniastrum africanum, Khaya ivorensis, Nauclea diderrichii,* species of the *Mitragyna* genus, *Dumoria heckelii* and *Tarrietia utilis.*

48

Fig. 27. Semi-deciduous moist tropical forest of the *Celtis–Triplochiton* association. Surroundings of Lake Bosumtwi, Ghana.

Trees of the middle storey attain a height of 7–40 m. These are mostly tree species which are less utilized on the local market, such as species of the *Berlinia* genus, *Allanblackia parviflora, Calpocalyx brevibracteatus, Cola chlamydantha, Funtumia africana, Pentadesma butyracea, Strombosia pustulata* and other tree species.

In the lower stratum of shrubs and herbs there occur species of the *Alchornea, Harrungana* and *Heisteria* genera, and *Bertiera racemosa, Mussaenda chippii, Randia hispida,* etc.

Although the evergreen tropical forest is very rich in tree species and diversified as to spatial order, it should be noted that it has a relatively small representation of the so-called African mahoganies, i.e. species of the *Entandrophragma* genus, the timber of which is especially sought after on world markets. According to data given by the Central Forest Administration in Ghana the surface area of the evergreen tropical forest is 752,395 ha.

The transition zone of the tropical rain forest in the *Lophira–Triplochiton* association adjoins to the south the previous formation and forms a 15–50 km wide belt; physiognomically its distinctions are slight. Gradually to the north, tree species appear in these forests which in the dry season shed their foliage for a short time, e.g. species of the *Bombax* and *Celtis* genera among others. The characteristic tree

49

species of this ecotype of the tropical forest include the genera *Lophira*, *Tarrietia*, *Cynometra* and especially *Triplochiton* and *Celtis*.

As to log volume in the *Lophira–Triplochiton* association the most represented commercial tree species are *Piptadeniastrum africanum*, *Triplochiton scleroxylon*, *Nesogordonia papaverifera*, *Antiaris africana*, *Entandrophragma cylindricum*, *E. angolense*, *Khaya ivorensis*, *Lophira alata*, *Chlorophora excelsa* and other species. Less frequent are, e.g. *Parkia bicolor*, *Pterygota macrocarpa*, *Cola gigantea* and *Tarrietia utilis*, while *Afrormosia elata* and *Guibourtia ehie* are absent. More frequent towards north are species of the *Entandrophragma* and *Khaya* genera, *Lovoa trichilioides* and *Antiaris africana*. In the lower tree storeys *Cola chlamydantha*, *Hymenostegia afzelii*, *Funtumia africana*, *Protomegabaria staphiana* and other tree species occur. The transition zone of the evergreen tropical forest covers a surface area of 840,450 ha.

The semi-deciduous moist tropical forest of the *Celtis–Triplochiton* association amounts to about 48% of the area of the high tropical forest in Ghana.

In this ecotype species of the *Celtis* genus, e.g. *C. adolfi-frederici*, *C. zenkeri*, *C. milbraedii* and *Triplochiton scleroxylon* occur frequently. Also well represented are *Khaya anthotheca*, *Streculia oblonga* and *Cylicodiscus gabunensis*. *Lovoa trichilioides* and *Cynometra ananta* are of rarer occurrence, while *Tarrietia utilis* is absent.

As to log volume in this association the following commercial tree species are represented most frequently: *Piptadeniastrum africanum*, *Antiaris africana*, *Triplochiton scleroxylon*, *Nesogordonia papaverifera*, *Turraeanthus africanus*, *Mansonia altissima*, *Guarea cedrata*, *Lophira alata*, *Entandrophragma cylindricum*, *Khaya ivorensis*, *K. anthotheca*, *Entandrophragma candollei*, *E. utile*, *E. angolense*, *Chlorophora excelsa*, *Terminalia ivorensis*, etc.

In the lower tree storeys the following species are frequent: *Funtumia elastica*, *Corynantha pachyceras*, *Hymenostegia afzelii*, *Monodora myristica*, *Sterculia tragacantha*, species of the *Myrianthus* genus and many other species. The *Celtis–Triplochiton* association covers an area of 3,931,620 ha.

The deciduous moist tropical forests of the *Antiaris–Chlorophora* association occur in the northwestern and northeastern regions of the high tropical forest, and in the mountainous border region with Togo. This formation of the high tropical forest, compared to others, has a less abundant representation of tree species and – on its northern border – a less luxuriant growth.

In the upper tree storey species of the *Celtis* genus and *Triplochiton scleroxylon* occur. There are also good growing conditions for *Antiaris africana* and *Chlorophora excelsa*. In the western regions the very valuable *Afrormosia elata* occurs. As to volume in the *Antiaris–Chlorophora* association the following tree species are represented most frequently: *Nesogordonia papaverifera*, *Afrormosia elata*, *Chlorophora excelsa*, *Antiaris africana*, *Piptadeniastrum africanum*, *Mansonia altissima*, *Entandrophragma utile*, *E. cylindricum*, *E. angolense*, *Guarea cedrata*, *Khaya anthotheca*, *K. grandifoliola*, *Terminalia ivorensis* and other tree species.

Fig. 28. Tropical forest giving way to agricultural land. Region of Kumasi, Ghana.

In the lower tree storeys the same tree species occur which are usually found in the previously described forest association. There are in addition species such as *Chidlowia sanguinea,* and on the border with Togo *Funtumia africana* occurs.

The formation of the closed tropical forest of the *Antiaris–Chlorophora* association occurs over an area of 2,701,370 ha, of which 21% is managed by the state (Malm, 1974).

The original untouched area of closed tropical high forest in Ghana covered 8,225,840 ha. However, this area now includes 1,557,108 ha of state managed forests and Forest Reserves (19%) and 539,756 ha (6%) of other forests. On the remaining area of 6,128,976 ha (75%) of the original region of closed forest, there are cocoa plantations, agricultural fields and the remains of devastated forests (Fig. 28).

The area of the actual natural closed high forest is 2,096,864 ha, i.e. one quarter of the total area of the original closed forest. Of the country's total surface area, the existing closed tropical high forest covers only 8.8% (Richardson, 1969).

To the north and southeast of the region of closed tropical high forest there is a vast extension of savanna woodland. This is a grass formation covered sporadically by low, sparse trees and bushes.

On the southern border belt of the savanna woodland there also occur tree species of the moist deciduous forest of the *Antiaris–Chlorophora* association. Northwards these species gradually disappear. Other tree species of a poorer growing capacity make their appearance. Examples include: on relatively more fertile land *Burkea*

*africana, Daniellia oliveri, Hannoa undulata, Hymenocardia acida, Lophira lan-
ceolata, Vitex cuneata;* in river valleys and their neighbourhoods, among the more
valuable species are *Afzelia africana, Celtis integrifolia, Khaya senegalensis* and
Pterocarpus erinaceus, and among other species, *Acacia campylacantha, Kigelia
aethiopica, Berlinia grandifoliola, Cynometra vogelii* and *Mimosa nigra* occur.

On extensive drier sites, mainly in the northern regions, *Adansonia digitata,
Acacia albida, Tamarindus indica,* species of the *Parkia* genus, shrubs of the *Combretum* genus and other species are of common occurrence.

The woodland, tree and shrub savanna regions extend over an area of about 15
million ha, equalling 66% of Ghana's territory. The actual area of open forests is
approximately 9.7 million ha, i.e. about 62% of the whole region of woodland and
tree savannas. The area of forests managed by the state in this dry region amounts
to 880,000 ha, or 9% of the area of dry open forests.

The coastal belt of bush and savanna extends from the port of Takoradi outwards
to the east. In the western part of this region shrub formations prevail, to the northeast of Accra grass land and shrub savanna extend.

The following tree species occur most frequently in the western part of the shrub
formation: *Baphia nitida, Dichapetalum barteri, Hymenostegia afzelii, Sophora occidentalis,* and as for more valuable species there are, e.g., *Antiaris africana, Ceiba
pentandra, Sterculia tragacantha,* the palm *Phoenix reclinata* and other tree species.

In the eastern part of the maritime region characteristic shrub species of the
Abutilon genus, as well as *Allophylus warneckei, Grevia carpinifolia, Elaeis drupi-
fera* and the introduced *Mangifera indica* occur. It should also be noted that these
maritime vegetational formations are gradually being turned into agricultural land.
Coastal bush and savanna extend over an area of about 450 thousand ha.

Mangrove forests occur in Ghana mainly in the estuary of the Volta river, but
also elsewhere along the coast where the surf does not break upon the shore. These
forests grow on an area of about 128 thousand ha. Only a few tree species are represented here. Species of the *Avicennia, Rhizophora* and *Laguncularia* genera are
typical.

4.1.3 Main economic tree species

Three hundred and sixty tree species which grow to a diameter of 70 cm and more
are classified in Ghana into four classes according to their economic importance and
utilization (Fergusson, 1969).

The first class includes commercial tree species of primary importance, subdivided into groups A, B and C.
(A) *Chlorophora excelsa* BENTH. et HOOK. f. – iroko, odoum
    Entandrophragma angolense C. DC. – tiama
    E. cylindricum SPRAGUE – sapele
    Khaya anthotheca C. DC. – African mahogany

K. grandifoliola C. DC. – African mahogany, bigleaf khaya

K. ivorensis A. CHEV. – African mahogany, red khaya

Dumoria heckelii A. CHEV. – makoré, cherry mahogany

Nauclea diderrichii MERR. – opepe, bilinga

(B) *Afrormosia elata* HARMS – kokrodua, afrormosia, asamela

   Lovoa trichilioides HARMS – African walnut, dibetou

   Terminalia ivorensis A. CHEV. – idigbo, black afara, emri

   Triplochiton scleroxylon K. SCHUM. – wawa, obeche

(C) *Tarrietia utilis* SPRAGUE – niangon, nyanhom

The second class includes tree species which are also economically important; they are used mainly for the domestic market and partially exported; this class also is subdivided into groups A and B:

(A) *Entandrophragma candollei* HARMS – omu, kosipo, heavy sapele

   Guarea cedrata PELLEGR. – scented guarea, white guarea

   G. thompsonii SPR. et HUTCH. – guarea, black guarea, diambi

   Lophira alata BANK ex GAERTN. f. – ekki, red ironwood

   Piptadeniastrum africanum BRENAN – dahoma, dabema

(B) *Antiaris africana* ENGL. – antiaris, bonkonko

   Guibourtia ehie J. LÉONARD – ovangkol, amazakoué

   Mansonia altissima A. CHEV. – mansonia, bété

   Mitragyna stipulosa O. KUNZE – abura

   M. ciliata AUBR. et PELLEGR. – abura

   Nesogordonia papaverifera R. CAPURON – danta

   Turraeanthus africana PELLEGR. – avodiré

The third class comprises tree species which are intended to be used by industry in the near future; they include 23 species, e.g.

   Afzelia africana SMITH – afzelia, apa

   Albizia ferruginea BENTH. – West African albizia, tanga-tanga

   Anopyxis klaineana ENGL. – bodioa

   Celtis zenkeri ENGL. – esa, African celtis

   Combretodendron africanum EXELL – essia, wulo

   Cynometra ananta HUTCH. et DALZ. – apomé

   Pycnanthus angolensis WARB. – ilomba, pycnanthus, acomu

   Terminalia superba ENGL. et DIELS – white afara, limba

The fourth class includes 311 tree species which in Ghana at present are of no economic importance. Among them are *Alstonia boonei*, *Aningeria altissima*, *A. robusta*, species of the *Berlinia*, *Bombax*, *Cola*, *Chrysophyllum* and *Daniellia* genera, *Ceiba pentandra*, *Chlorophora regia*, *Coelocaryon oxycarpum*, *Fagara macrophylla*, *Klainedoxa gabonensis* var. *oblongifolia*, *Maesopsis eminii*, *Manilkara lacera*, *Ongokea gore*, *Parinari excelsa*, *Pentaclethra macrophylla*, species of the *Pterygota* genus, *Ricinodendron heudelottii*, species of the *Uapaca* and *Vitex* genera and other tree species which, in other countries, are used in industry and sold on local markets or even exported, such as *Aningeria altissima*, species of the *Berlinia* genus, *Ceiba pentandra, Chlorophora regia, Fagara macrophylla, Maesopsis eminii,* species of the *Pterygota* genus, etc. (Thompson, 1973).

The most important tree species of Ghana can be classified by the weight of air dry timber into the following categories:

(A) Light timber up to 600 kg per m^3

Wawa or abachi or samba *(Triplochiton scleroxylon)* – soft timber of light yellow colour; used for plywood, packing material and light furniture; it is the most exported timber from Ghana.

Fromager or ceiba or kapok *(Ceiba pentandra)* – rather strong timber of light reddish-grey colour; used for plywood, boat construction, packing material and insulation.

Alstonia or sindru *(Alstonia boonei)* – soft timber of yellowish-white colour, suitable for packing materials, plywood, etc.

Essessang or ricinodendron *(Ricinodendron heudelottii)* – soft timber of yellowish-grey colour, suitable for pulp production; used also for packing materials and insulation.

(B) Medium-heavy timber of 600 to 800 kg per m^3

Khaya or acajou d'Afrique *(Khaya* spp.) – of light reddish to light russet colour, a much sought-after timber for heavy furniture, decorative veneer and plywood.

Sapele or "banded mahogany" *(Entandrophragma cylindricum)* – timber of reddish-brown colour; used mainly for decorative veneer and heavy furniture; one of the most desired timbers on the world market.

Sipo or utile *(Entandrophragma utile)* – of reddish-brown colour, darker than sapele; exported in the form of logs and sawn timber. Suitable for decorative veneer, furniture, carpentry.

Lovoa or African walnut *(Lovoa trichilioides)* – of light brown colour, very decorative; used for heavy furniture, in carpentry, as construction timber and mainly for veneer and plywood.

Avodiré *(Turraeanthus africana)* – less durable timber of lemon-yellow colour; used for surface and interior veneer, in carpentry, wood-carving and for packing material.

Aiélé or canarium *(Canarium schweinfurthii)* – timber of pale rosy-grey colour; used for decorative veneer, in carpentry, for massive furniture and panelling.

Framiré or emeri *(Terminalia ivorensis)* – less durable timber of light yellowish-brown colour; used in construction, parquetry, for cheaper furniture, for veneer and plywood.

Antiaris or ako *(Antiaris africana)* – non-durable timber of yellowish-grey colour; used for plywood, blockboard, crating, light construction and in carpentry.

Ilomba or otié *(Pycnanthus angolensis)* – timber of light brown to rosy-brown colour, resembling to okoumé; used for plywood, panelling, matches, cheaper furniture and for packing material.

(C) Heavy timber of 800 to 1,000 kg per m^3

Tiama or edinam *(Entandrophragma angolense)* – timber of russet colour; used as other African mahoganies for heavy furniture, decorative veneer and in parquetry and construction.

Kosipo or omu *(Entandrophragma candollei)* – timber of dark brownish-red to reddish mauve colour, durable; used in building construction, parquetry, for flooring, furniture and veneer.

Makoré or baku *(Dumoria heckelii)* – durable timber of russet colour; used mainly for decorative veneer, plywood, in parquetry and carving.

Mansonia or bété *(Mansonia altissima)* – very decorative timber, dark red to mauve; used for decorative veneer, luxury furniture, pianos and panelling.

Niangon or nyanhom *(Tarrietia utilis)* – decorative russet timber; used in carpentry, parquetry, for furniture making and decorative veneer.

Iroko or odoum *(Chlorophora excelsa)* – very durable timber of dark yellow to brown, one of the most valuable timbers in Africa; used in building construction, ship building, bridge construction, cabinet-making, furniture-making and carving.

Padouk or barwood *(Pterocarpus soyauxii)* – durable and decorative timber of reddish colour; used in carpentry and carving, for bridge construction, railway sleepers and decorative veneer.

Abura or subaha *(Mitragyna spp.)* – less durable timber, russet to redish-grey; used for plywood, cheap furniture, in carving and carpentry.

Kotibé or danta *(Nesogordonia papaverifera)* – durable timber, russet; used for textile shuttles, gun butt-ends and golf clubs instead of hickory, for flooring, in parquetry and for decorative veneer replacing sapele.

Guarea or bossé *(Guarea spp.)* – timber of rosy brown colour resembling sapele, durable; used for heavy furniture, in parquetry, construction, carving and for decorative veneer.

Dabema or dahoma *(Piptadeniastrum africanum)* – a yellowish-brown timber resembling iroko by its colour, but not with interior mechanical and technological properties; used in construction, parquetry, railway sleepers, functional furniture and plywood.

Koto or pterygota or kyere *(Pterygota spp.)* – less durable timber of yellowish-grey colour; used for veneer and plywood, cheaper furniture, packing material and particle boards.

(D) Very heavy timber above 1,000 kg per m^3

Kokrodua or afrormosia or asamela *(Afrormosia elata)* – one of the most valuable timbers in Africa; it resembles teak by its yellowish-brown colour; used for heavy luxury furniture, decorative veneer, in harbour construction, parquetry and turnery.

Azobé or kaku or ekki or red ironwood *(Lophira alata)* – timber of a brownish-rosy to mauve colour. The heartwood is very durable and strong. It is used in heavy construction, harbour construction, parquetry, for railway sleepers and textile shuttles.

Bilinga or kusia or opepe *(Nauclea diderrichii)* – a decorative timber of lemon-yellow to golden-orange colour, resembling Indian satinwood. It is exported and used mainly for decorative veneer. In the countries of occurrence it is used for building and bridge construction, in parquetry, for flooring and in carving.

Afzelia or doussié *(Afzelia africana)* – very durable, hard timber of russet colour; used in construction including harbours, in parquetry, railway sleepers, furniture and decorative veneer.

Ovangkol or shedua *(Guibourtia ehie)* – a yellowish-brown timber with a russet hue. It is used mainly for decorative veneer replacing walnut, for panelling and decorative flooring.

4.1.4 Forest inventory

The rational utilization of tropical forests in Ghana required basic information on their tree species composition and diameter structure. Therefore steps were taken for a partial inventory of forests managed by the state according to the principles of mathematical statistics. This was basically an objective selection; the inventory being limited to a part of the whole forest area with the sample units – inventoried strips – distributed systematically over the relevant forest area (systematic sampling design).

For forest inventory a sampling intensity of 5% was established. On each square mile (= 259 ha) two inventory strips had to be surveyed, each one measuring 1,600 × 40 m. The original decision to identify and record all tree species in the inventoried plots with a diameter larger than 10 cm was not observed. Neither was the sampling intensity of 5% adhered to. In spite of that, very valuable information on the condition and composition of state managed forests in Ghana was obtained by this national inventory.

For the determination of the growing stock in the inventoried forests no volume tariff tables were set up for the various tree species or their groups, but the following conversion factors (in English imperial measurements) were used:

Girth class (feet)	Basal area (Hoppus square feet)
3–5	1.0
5–7	2.2
7–9	4.0
9–11	6.25
11–13	9.0
13–15	12.25
15+	16.0

The conversion factors of the basal area to the stem volume were as follows:

Non-exploitable trees to the girth of 9 feet, (i.e. about D.B.H. = 90 cm), 1 Hoppus square foot = 30 cubic feet ($\doteq$ 0.85 m^3).

Exploitable trees with a girth exceeding 9 feet, 1 Hoppus square foot = 44 cubic feet ($\doteq$ 1.25 m^3).

The national inventory of high forests in forest reserves over an area of about 1.5 million ha was terminated in 1970.

According to the publication "Ghana Forest Products Transport Study" (Dawson et al., 1972) the growing stock of commercial tree species of the first and second class (including 25 tree species) with a diameter (D.B.H.) exceeding 70 cm amounted to 90 million m^3 in the Ghanian high forests managed by the state (1970). The growing stock of tree species of the third class (including 23 tree species) represented another 25 million m^3 of timber. Economically important tree species of the first to third class were thus amounting to a growing stock of 115 million m^3. From this timber stock, 43 million m^3 (34%) were formed by trees with a diameter exceeding 110 cm.

It follows that in the Ghanian forest reserves at this time, the average growing stock of mature timber of economic tree species of the first to third class (together 48 tree species) amounted to 77 m^3 per hectare. The most represented tree species of the growing stock of mature trees (D.B.H. $\geq$ 70 cm) in the inventoried region are given in Table 9. The tree species mentioned amounted to 74% of the total growing stock of economic tree species in forests managed by the state.

Table 9. Tree species with the largest exploitable growing stock in Ghana

Tree species	Growing stock	
	10^6 m^3	%
Triplochiton scleroxylon	20.2	17.6
Piptadeniastrum africanum	14.8	12.9
Antiaris africana	14.6	12.6
Nesogordonia papaverifera	6.8	5.9
Mansonia altissima	4.4	3.8
Turraeanthus africana	4.2	3.6
Lophira alata	3.0	2.6
Entandrophragma cylindricum	2.9	2.5
Khaya spp.	2.8	2.4
Chlorophora excelsa	2.7	2.3
Guarea spp.	2.7	2.3
Entandrophragma utile	1.8	1.6
Afrormosia elata	1.7	1.5
Entandrophragma candollei	1.6	1.4
Entandrophragma angolense	1.3	1.1

In 1970 a total of 1,920 thousand m^3 of industrial wood from tree species of the first, second and third class was exploited, i.e. not even 0.2% of the total growing stock of commercial tree species. On the assumption that the economic tree species are being naturally regenerated and that no degradation of forests takes place, the annual exploitation of industrial wood is not excessive with regard to the total growing stock in Ghana. Wawa is the most exploited and exported tree species in Ghana, in 1970 annual fellings of this tree species amounted to 642 thousand m³, representing 33% of annual fellings of all economic tree species and 3% of the total resources of mature trees of this species. The log extraction of African mahoganies, including species of the *Entandrophragma* and *Khaya* genera, was 530 thousand m³ in that year, representing 28% of fellings of all economic tree species. The fellings of African mahoganies amounted to 5% of the growing stock of mature trees (D.B.H. ≥ 70 cm). Thus timber resources will be exhausted in the course of 30–40 years if the necessary steps are not taken to obtain the safe reproduction of the above-mentioned commercial tree species.

In forests managed by the state, out of the growing stock of all economic tree species, 7% belonged to tree species of the first A class; 60% to tree species of the first B, C and second class; and 33% to tree species of the third class.

In the inventory of forests managed by the state trees were callipered beginning at a diameter of 30 cm (girth of 3 feet) and were classed into diameter classes with a span of 20 cm (girth class of 2 feet). Because the directive to calliper all trees in the forest was frequently not observed, the author used inventory data from selected forest sites where callipering included economic tree species of the first and second class.

For the theoretical expression of distribution of commercial tree species by their diameter breast height the author used the mathematical formula of the beta function in selected forest localities, namely in the region of moist evergreen forests (Subri Forest Reserve) and in the region of moist semi-deciduous tropical forest (Attandaso Forest Reserve, Wawahi Baku Forest Reserve, Bosumkese and Asukesi Forest Reserves and Tain Forest Reserve) (Borota, 1981). All these forests have been exploited in the past. Inventory data from different parts of Ghanian high tropical forests were used in order to demonstrate, on a concrete basis, the different diameter structure of trees and the heterogeneity of composition of commercial and other economically important tree species in selected forest reserves.

For testing the significance of the calculated values the χ^2 (chi-square) test of good agreement was used so that the resulting value was compared to the critical value of distribution χ^2 in the relevant degrees of freedom on the 5% level of significance. Kolmogorov–Smirnov's test was also used (Plokhinskii, 1937). It should be noted that the distributions of tree numbers of certain tree species did not correspond to the beta distribution. The final value fo the χ^2 test exceeded the critical tabulated value.

The forests of the Subri Forest Reserve are situated in the southwestern part of the country in the region of moist evergreen tropical forests which lie to the north of the port of Takoradi. The forest inventory covered 2.5% of the forest area, the trees were callipered beginning with a diameter breast height of 10 cm and 30 cm, respectively. As an illustration the diameter distribution of trees for the main commercial species, computed by the beta function, is graphically shown in Fig. 29. Compensated values of the diameter distribution of trees for certain tree species by the beta function are given in Table 10.

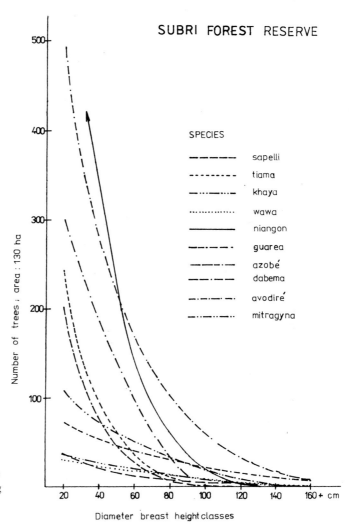

Fig. 29. Diameter distribution of trees for the main tree species according to the beta function. Subri Forest Reserve, Ghana.

Niangon, dabema and avodiré are the most frequently represented commercial tree species in the region of the Subri Forest District. These tree species are typical for the moist evergreen African tropical forests. Niangon with a diameter breast height (D.B.H.) ≥ 30 cm was represented by 17.1 trees per hectare. The representation of dabema was 8.9 trees and the average number of avodiré trees was 5.0 trees per hectare, always for D.B.H. ≥ 30 cm.

Table 10. Theoretical data of the beta-diameter distribution. Some timber species of Subri Forest Reserve

Midpoint value of D.B.H. class	Timber species				
	Sapele	Tiama	Khaya	Niangon	Dabema
cm	Number of trees, area: 130 ha				
20	38	–	–	–	–
40	19	62	31	311	225
60	12	27	30	157	149
80	8	13	23	66	101
100	6	6	15	23	66
120	4	2	8	2	39
140	2	1	4	–	19
160+	1	–	1	–	4
Total	90	111	112	559	603
Const.	0.0007	0.0000	0.0001	0.0001	0.0084
α	−0.4472	−0.3270	0.2595	−0.0626	−0.1348
β	1,6840	3.0501	2.0058	2.6982	1.5416
χ^2	6.097	4.209	1.920	5.501	13.168

Among commercial tree species of the *Entandrophragma* genus tiama *(E. angolense)* had the largest representation, while *Khaya ivorensis* occurred less frequently. This tree species, much sought after on world markets, has been practically exhausted in the coastal forests of Ghana.

The Attandaso and Wawahi Baku Forest Reserves are situated in the southern part of the country, in the transition belt between the moist evergreen and the moist semi-deciduous tropical forests. In the Attandaso Forest Reserve the forest inventory covered 5%, in the Wawahi Baku Forest Reserve it covered 1% of the forest surface (Addo and Baidoe, 1962).

In the Attandaso Forest Reserve dabema is represented most frequently, with 6.3 trees per hectare for D.B.H. ≥ 10 cm. Niangon, azobé and avodiré do not occur. Kotibé with 5.8 trees per hectare are frequent. Among African mahoganies, tiama is the most represented, with 2.6 trees per hectare for D.B.H. ≥ 10 cm. It is followed by sapele with 1.7 trees per hectare, and khaya with 1.3 trees per hectare. In the Attandaso Forest Reserve the representation of mature trees of these three species was 6 trees per hectare for D.B.H. ≥ 90 cm.

The region of the Wawahi Baku Forest Reserve is situated not far off to the northeast from Attandaso. Here diambi is the most frequent tree species, with 12.3 trees per hectare. This lesser-known tree species attains diameter breast height of 100–120 cm.

Table 11. Actual data of diameter distribution of some timber species. Wawahi-Baku Forest Reserve, area: 259 ha

Species	Midpoint value of D.B.H. class in cm								
	20	40	60	80	100	120	140	160+	Total
Entandrophragma cylindricum (sapele)	387	111	94	84	94	36	17	–	823
E. angolense (tiama)	894	161	47	20	18	6	8	1	1,155
Khaya ivorensis (khaya)	261	58	57	20	17	11	1	–	425
Mansonia altissima (bété, mansonia)	233	167	121	36	3	–	–	–	560
Guarea cedrata (guarea)	2,876	207	69	17	3	–	–	–	3,172
Nesogordonia papaverifera (danta)	1,468	688	370	113	20	–	–	–	2,659
Triplochiton scleroxylon (wawa, obeche)	367	224	264	291	163	70	20	3	1,402
Antiaris africana (ako, antiaris)	–	178	101	158	128	19	10	–	594
Piptadeniastrum africanum (dabema)	2,077	241	121	98	57	26	10	1	2,631

Among other commercial tree species kotibé is frequent (10.2 trees per hectare for D.B.H. ≥ 10 cm) attaining a diameter breast height of up to 100 cm, and dabema is likewise well represented (10.1 trees per hectare for D.B.H. ≥ 10 cm). Among mahoganies, tiama occurs most frequently, with 4.5 trees per hectare for D.B.H. ≥ 10 cm. Sapele is also important, with 3.2 trees per hectare for D.B.H. ≥ 10 cm. *Khaya ivorensis* was represented with 1.6 trees per hectare for D.B.H. ≥ 10 cm. For a clearer view, empirical values of the diameter distribution of trees of some commercial species are given in Table 11.

The Bosumkese and Asukesi Forests are situated in the Brong Ahafo region, to the south of the town of Sunyani, in the western part of Ghana. They are affiliated to the association of *Antiaris–Chlorophora* tree species. The Asukesi Forest is situated to the west of Bosumkese. In both forest reserves the inventory covered 3% of the forest area.

According to inventory data, in the Bosumkese Forest Reserve wawa is the most frequent commercial tree species with 11 trees per hectare for the diameter breast height (D.B.H.) ≥ 10 cm, and 1.5 trees per hectare for D.B.H. ≥ 90 cm. It is followed by kotibé with 5.7 trees per hectare for D.B.H. ≥ 10 cm. Of less frequent occurrence is dabema, with 3.4 trees per hectare. Although these forests are botanically classified as belonging to the *Antiaris* and *Chlorophora* association, these two tree species are little represented (1.2 to 1.4 trees per hectare for D.B.H ≥ 10 cm). From among mahoganies tiama is the most frequent, with 2.8 trees per hectare; sapele, sipo and khaya are represented almost equally, with 1.0 to 1.1 trees per hectare (Baidoe, 1964).

In the Asukesi Forest Reserve wawa is the most frequent commercial tree species, with 16.1 trees per hectare for D.B.H. ≥ 10 cm, and 1.9 trees per hectare for D.B.H. ≥ 90 cm. From among other tree species kotibé, with 4.2 trees per hectare for D.B.H. ≥ 10 cm, and bété, with 3.8 trees per hectare, are most common. Among mahoganies, sapele, with 3.3 trees per hectare for D.B.H. ≥ 10 cm, has the largest representation. From among all the regions investigated in Ghana, sapele is most common in the Asukesi Forest Reserve. Sipo occurs less frequently, with 1.6 trees per hectare. In the Asukesi and Bosumkese Forest Reserves *Khaya invorensis* which is absent, is replaced by *Khaya anthotheca* which, in both forest reserves, is represented approximately by one tree per hectare.

In the Asukesi Forest Reserve kokrodua, one of the most valuable tree species in western Africa, is important, with an average representation of 2 trees per hectare for D.B.H. ≥ 10 cm, and 0.8 trees per hectare for D.B.H ≥ 90 cm.

For the sake of comparison, the compensated values of the diameter distribution of trees according to the beta function are shown in Fig. 30 for the aggregation of all commercial tree species in the moist semi-deciduous tropical forests in the region of the Wawahi Baku, Attandaso, Bosumkese and Asukesi Forest Reserves.

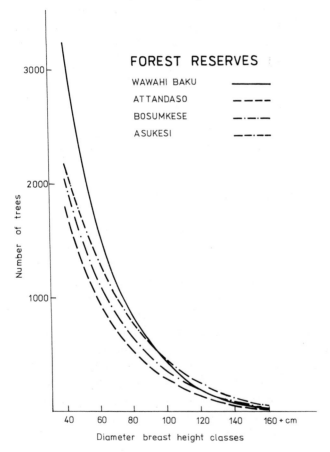

Fig. 30. Diameter distribution of trees for certain commercial treee species according to the beta function in selected localities in Ghana.

The Tain Forest Reserve extends in the Brong Ahafo region along the northwestern border of the belt of high moist tropical forest. This forest region is less abundant in tree species, and of poorer growth. The forest inventory covered a forest surface of 624 hectares with a 1% selection. Only five important commercial tree species were callipered, beginning with a diameter breast height of 30 cm. The most common commercial tree species, wawa, was represented on average by 4.1 trees per hectare for D.B.H. $\geq$ 30 cm. It is followed by iroko with an average number of two trees per hectare. From among other tree species, mansonia and kokrodua are of the greatest importance, with the occurrence of about one tree per hectare for the diameter breast height exceeding 30 cm. In the Tain Forest Reserve the trees attain up to 120 cm diameter breast height (Andoh et al., 1965).

4.1.5 Timber processing and export

Timber logging in natural tropical forests has its own peculiarities. In the heterogeneous tropical high forests with the abounding occurrence of different tree species, only commercial, or other economically important tree species (frequently of enormous dimensions with a volume of 20 m^3 or more) of industrial wood are exploited. Many tree species in the forest remain untouched by logging operations. There is no interest in their timber, neither on the domestic, nor on foreign markets.

Timber exploitation for industrial consumption and timber skidding by automative mechanization equipment usually cause heavy damage to forest stands. Caterpillar crawler and wheeled tractors equipped with winches for hauling cables and other facilities are being used (Figs 31, 32). Any such logging operation in the natural tropical forest impairs the ecological balance and devaluates the original forest. Some years after logging a secondary forest, or second-growth comes into being which is deprived of the most valuable tree species. The representation of undesired tree species, like species of the *Musanga, Macaranga* and *Trema* genera in tropical Africa, increases. Such a situation occurs in Ghana, too.

Forestry organized according to the principle of yield regulation could succeed in the renewal of tropical forests, too, by preserving their substance. The basic problem is to enrich the tropical forest with valuable tree species. In order to achieve the regeneration of tropical forests qualified forest personnel are required who would also be responsible for the control of weeds, lianas and inferior tree species.

In industrial timber logging, only the trunk wood is utilized, i.e. the volume of stems from the buttresses to the first branches of the tree crown. The trunk wood of mature and over mature trees may amount to 2–35 m^3. Wood waste remaining in the forest after logging is frequently more than 50% of the tree volume.

Closed high forests play an important role in the economy of Ghana. Timber exploitation in Ghana is regulated in agreement with the existing forest law.

Timber for industrial use may be felled only after obtaining a licence, i.e. an official permit for logging. At present such licences are issued in Ghana for a period

Fig. 31. Skidding of logs by tractor. Unum-Su Forest Reserve, Ghana.

Fig. 32. Skidding of logs by elephant. Nilambur, India.

of 20–30 years. For this an annual rent is paid according to the area of forest to be logged.

For the felled timber, special licence fees or "royalties" are paid. The amount of the fee depends on the economic importance of the tree species. The highest fee was (in 1973) for kokrodua: 33.60 ₵ per tree (1 ₵edi = 0.870 US$ in 1977). For mahogany, including sapele and iroko 19.20 ₵, and for wawa 9.60 ₵ were paid. The same fee is paid for each tree of a certain timber species regardless of whether the diameter is 80 or 160 cm, or whether the length of the trunk is 10 or 30 m.

The allowable cut is established every year by the forest administration on the basis of results of a 100% inventory of commercial species which only begins at a certain logging girth (diameter breast height) of trees.

It is estimated that in Ghana's forests there are about 150 timber species which are, or could be in the near future, of importance in industrial processing. Up to now in Ghana about 30 tree species have been of commercial importance.

As a point of interest, the recorded felling of stems and the average volume of stems of the 20 most exploited commercial tree species in Ghana in 1970 are listed in Table 12.

Table 12. Recorded felling of stems and the mean stem volume of the 20 most exploited commercial tree species in Ghana in 1970 (Dawson et al., 1972)

Timber species	Volume of industrial wood removals	Volume of mean stem
	m^3	
Triplochiton scleroxylon (wawa)	641,889	25.0
Entandrophragma cylindricum (sapele)	172,173	14.2
Entandrophragma utile (sipo)	126,821	22.7
Khaya ivorensis (khaya)	116,755	20.0
Afrormosia elata (kokrodua)	97,139	15.3
Chlorophora excelsa (iroko)	69,174	18.8
Entandrophragma angolense (tiama)	60,048	18.1
Guibourtia ehie (ovangkol)	48,257	13.1
Dumoria heckelii (makoré)	45,349	19.8
Mansonia altissima (bété)	29,906	4.0
Khaya grandifoliola (khaya)	23,047	18.1
Khaya anthotheca (khaya)	22,272	19.7
Terminalia ivorensis (framiré)	17,610	16.7
Piptadeniastrum africanum (dabema)	15,688	7.4
Cylicodiscus gabunensis (okan)	10,445	11.9
Entandrophragma candollei (kosipo)	8,649	13.7
Antiaris africana (ako)	6,609	8.5
Tarrietia utilis (niangon)	6,211	2.9
Nesogordonia papaverifera (danta)	5,892	2.2
Celtis spp. (celtis)	5,332	4.0
Total	1,529,266	16.5

These 20 tree species were shared very unevenly in the annual felling of industrial wood. About 60% of all logging comprised three tree species: wawa, sapele and sipo. The share of wawa in annual log production was 41%.

The average log volume was computed from the actual number of felled trees. The length of the felled logs was 15–30 m. The average volume of felled trees indicates that there is a high timber volume of the commercial and less important tree species of the Ghanian tropical high forests. At present lesser-known tree species are not utilized.

Until 1975 more than half of the annual production of industrial wood originated in forests not managed by the state. Now exploitation is shifting gradually to forests managed by the state.

Forest industries in Ghana are preferred components of the national economy. The basic element of the forest industry is timber logging which is mostly in the hands of logging companies. Firms own either smaller or larger sawmills and at the same time hold licences for the export of logs and sawn wood. Such firms are annually felling 75% of the logs felled in Ghana.

The felled trees are transported in the form of logs either to the sawmills and other woodworking plants, or directly to the port of Takoradi for export.

The most developed branch of the forest industries is sawmilling. In 1984 there were 65 sawmills in Ghana, half of them in the Ashanti region. The centre of sawmilling is the town of Kumasi.

The largest timber companies – today nationalized – are well equipped and well organized. These firms process part of the felled trees into sawn wood, and also into veneer and plywood, or dimension stock for furniture as the case may be. Sawn wood is used mostly in construction and furniture-making. Local consumption is relatively low. Taking into account the low wage level the price of forest products is rather high.

Ghana was one of the largest exporters of tropical timber in Africa. Timber exported from Ghana has good technological properties and is widely sought after on world markets, mainly in Europe.

Notwithstanding the tendency of the Ghanian authorities to limit the export of logs as a basic raw material and to increase the export of semi-finished articles and finished forest products, the share of logs in the export of forest products is still high (Table 13).

The export of forest products from Ghana, mainly of good quality logs, exceeded 1 million m^3 in 1960. After the establishment of a special state institution for timber, the Ghana Timber Marketing Board (G.T.M.B.) in 1961, which determines the minimum price for exported timber, the export of logs declined. In consequence of an increased demand for tropical timber in world markets the export of logs attained its maximum in 1973 with 1,387,000 m^3. But after the large price increase for tropical

Table 13. Export of the main forest products from Ghana in 1,000 m^3 (Yearbook of Forest Products, FAO, Rome, 1960, 1975, 1986)

	1960	1965	1970	1975	1980	1985
Logs	1,042	560	601	400	105	130
Sawn wood	236	230	241	165	69	80
Veneer sheets	0.8	0.4	0.4	8	8	13
Plywood	4	12	22	18	2	1

wood at the end of 1973 the export of timber from Ghana declined markedly during the next few years (see Table 13).

From among the large number of tree species occurring in the forests of Ghana only a few species are exported which in the world markets have a certain tradition based on their technological properties and timber colour.

Of the total export of logs in 1972 (951,000 m³) wawa had the largest share with 55% followed by sipo 14%, *Khaya* species 8%, sapele 7% and makoré 4%. These 5 tree species represented 88% of the total volume of logs export.

The share of tree species in the total export of sawn timber in 1972 was as follows: sapele 26.1%, wawa 14.8%, iroko 13.9%, sipo 13.2% and *Khaya* species 8.4%. These five tree species were responsible for 76% of exported sawn wood. On the domestic market sawn timber of iroko and wawa is the most required (Malm, 1974).

One aim of the Ghanian development plan for the years 1976–1980 was, among others, to achieve the effective utilization of all resources of the country and to increase her economic independence. The Ghanian Government also tries to solve the country's gloomy economic situation by a more effective utilization of timber resources. The maximum processing of logs in Ghana and increased exports of finished forest products are envisaged. With this intention the Government prohibited, with effect from 1st January 1979, the export of logs of the following commercial species: khaya, sapele, sipo, kosipo, tiama, makoré, mansonia, kokrodua, iroko, dibetou, niangon, ovangkol, avodiré and introduced teak.

Although this government injunction against log export cannot be considered nowadays as absolutely binding (since regulations are frequently changing according to the situation on world markets), the tendency towards enforcing the processing of raw timber in the country of origin remains obvious.

4.2 Republic of Congo

4.2.1 Basic data

The Republic of Congo (République Populaire du Congo) is situated in west-central Africa lying on both sides of the equator between 3°50′ N and 5°05′ S latitudes and between 11°05′ and 18°20′ E longitudes. It borders on Gabon in the west, on Cameroon and the Central African Republic in the north, and on Zaire and on part of Angola (Cabinda), in the east and south. In the southwest it borders on the Atlantic ocean. Congo covers a surface area of 342,000 km² with about 1.8 million inhabitants (1985). The largest part of the population inhabits the south of the country between the port of Pointe Noire and the capital Brazzaville.

The administrative division of the republic includes 9 regions and 47 districts. Brazzaville is the capital and, at the same time, the largest city (about 440,000 inhabitants in 1985). Pointe Noire (about 220,000 inhabitants) is the second largest city in the country and its modern harbour is of great commercial importance.

French is the official language but "lingala" in the north of the country and "kikongo" in the south are national languages. The CFA franc is the monetary unit (1 French franc = 50 CFA francs).

The majority of the population of the Republic of Congo is engaged in traditional agriculture. Despite the small number of inhabitants the country does not produce enough food for her population. Manioc (cassava), peanuts, maize, millet, rice and bananas are the main crops. Among industrial crops sugar-cane, copra, cocoa and coffee are of economic importance.

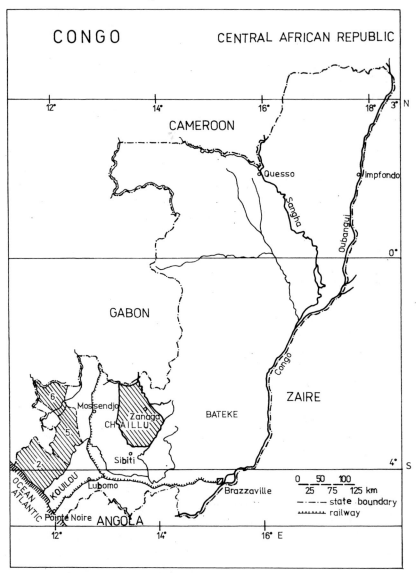

Fig. 33. Orientation map of the Republic of Congo.

Fig. 34. Landscape in the Niari region, Congo.

In the national economy of Congo oil extraction plays an important role (Elf Congo, Hydro-Congo, Agip). In the processing industry, forest industries concentrated in the port of Pointe Noire are in the forefront. Food and textile industries are developing, too.

The surface of the country is unevenly undulated. Behind a coastal plain which is approximately 50 km wide the forested Mayombe mountain ridge, with its highest peak Mbamba reaching to 810 m above sea level stretches right to the border with Gabon. Farther inland there is the wide valley of the Niari river and behind it the forested Chaillu mountain ridge, with its highest peak Mont Birougou reaching to 900 m above sea level, stretches again to the border with Gabon. In the central parts of the republic, at altitudes of between 500 and 800 m, there are the savanna plateaus of Koukouya, Djambala, Nsah-Ngo and Mbé together with their associated river valleys, the highest mountain being Peak Lékéti which rises to 1,040 m above sea level.

In the northeastern part of the country, to the east of 16° E longitude, there stretches the extensive Congo basin – a region of forests and swamps at an average altitude of 300 m. This region is drained by the important rivers of Oubangui, Sangha and Likouala. The northwestern part of the country is hill with swampy valleys and the isolated mountain of Nabemba which lies 1,000 m above sea level.

Fig. 35. Waterfalls on the Foulkari river, Congo.

Fig. 36. Liana bridge on the Ogooué river. Region of Zanaga, Congo.

70

The major part of the country relates to the drainage area of the Congo river, and of its tributary the Oubangui. Both rivers are navigable. The Congo river is navigable as far as the capital Brazzaville. The coastal region of the country is situated mainly in the basin of the Kouilou river which is navigable in its lower course (see Fig. 33 – the orientation map of Congo) (Figs 34–36).

The climate is tropical, equatorial, warm and moist, with an average annual temperature of 24–26°C, and slight daily and annual fluctuations. Annual precipitation varies from 1,300 to 2,000 mm, occurring mainly from February to April and from October to November. The relative air humidity does not sink below 70%. In the northern Congo, in the Ouesso and Impfondo regions the annual precipitation is 1,500–2,000 mm, the average annual temperature is 25–27°C, and the relative air humidity is 75 to 85%. In the southern parts of the country the dry season lasts for 3–4 months from June to September, in the north of the country it rains all year round.

4.2.2 Forests and forestry

The area of forests in the Republic of Congo is about 20 million ha. The tropical, evergreen or semi-deciduous closed high forests turn locally into bush savanna and grassland. There is abundant rainfall spread over the year except for a short period of drought, and humidity is very high. Such conditions afford a most suitable environment for forest growth (Aubréville, 1985).

The Congolese evergreen and semi-deciduous tropical forests are characterized both by a rich tree species composition and a varied spatial structure. The tallest trees attain a height of 50–60 m and a diameter breast height of 200 cm. On poorer sites the trees reach heights of 30–35 m. The herb cover and grasses grow in places from 2 to 3 m (Gilbert, 1984).

The tropical forests of Congo may be classified as follows:
- Tropical evergreen natural forests on solid soil; very valuable tree species occur here, including: *Aucoumea klaineana, Terminalia superba*, species of the *Entandrophragma* genus, *Lovoa trichilioides, Baillonella toxisperma* and other species.
- In northern Congo the forests cover 13 million ha of land where seasonally flooded or swampy forests grow over an area of about 7 million hectares. In these forests, trees frequently form buttresses or support roots (Fig. 37). Here, *Gilbertiodendron dewevrei* and species of the *Guibourtia* and *Uapaca* genera occur frequently.
- Degraded forests occur mostly in the vicinity of settlements and alongside the roads. They are the result of shifting cultivation. Typical tree species of such forests are *Musanga cecropioides* and species of the *Fagara, Macaranga* and *Vernonia* genera (Fig. 38).

All forests of Congo are property of the state. Their rich growing stock is one of the chief natural resources and timber exploitation is one of the main sources of

Fig. 37. Trees of the *Uapaca* genus form big buttresses. Region of Zanaga, Congo.

Fig. 38. Degraded tropical forest. Region of Makabana, Congo.

revenue of the state which contributes significantly to the development of the national economy.

Congo's forests are divided into three geographical regions which are:
- the Kouilou-Mayombe massif,
- the Chaillu-Niari massif,
- the massif of northern Congo.

The Kouilou-Mayombe forest massif is the natural barrier between the coastal zone and the interior of the country. Because of easy access, the forests of this region, rich in the commercial tree species limba and okoumé, have been intensely exploited since the World War II. The forests of this region have a surface area of over 1.2 million ha.

The Chaillu-Niari forest massif is situated on the border with southeastern Gabon. These forests where okoumé predominates cover an area of about 3.3 million ha. This forest region is also accessible thus allowing for the exploitation of timber resources. As a precondition for the rational utilization of the forests of this region, forest inventories were taken there in the early 1970s (Groulez, 1963).

The country's main timber resources are found in its northern part. There, the forests are much less impaired by logging than in the south. Forest inventories carried out in the regions of Sangha (1972) and Leikouala (1976) established the occurrence of the following tree species in northern Congo: sapele, limba, ayous (wawa), wengé, padouk, azobé, ilomba, bubinga, utile, kosipo, tiama, acajou (khaya) and other tree species. Logs extracted for export from Congo's northern region are transported on the Oubangui and Congo rivers to the capital Brazzaville and from there by the railway to the port of Pointe Noire.

The Ministry of Forestry is Congo's supreme authority for forestry affairs. There are provincial forest administrations in all provinces of the country, and in some districts there are district forest administrations.

The Ministry of Forestry supervises the Office Congolais de Forêts (OCF), a state organization entrusted with the afforestation of savannas for the production of both industrial and fuelwood, and also with the regeneration or enrichment of natural forests especially with limba. Also under the Ministry's control are the state logging organizations responsible for the felling and, in part, the processing of logs of both the main commercial and the lesser-known tree species.

In 1978 the mixed Congolese–French organization for reforestation by the use of eucalypts and pines for industrial use was established, i.e. Unité d'Afforestation Industrielles du Congo (UAIC). This organization cultivates eucalypt-hybrids (e.g. *Eucalyptus* hybrid 12ABL, *E.* hybrid 12ABL × *E. grandis* and others) in the surroundings of the port of Pointe Noire. In 1987 23,000 ha of eucalypt plantations and 2,000 ha of *Pinus caribaea* were established here. The average rotation of the eucalypt plantations is seven years. The average timber production is 160–180 m^3 per hectare. Mature stands attain an average diameter breast height of 20–25 cm and a mean height of 23–25 m. The timber is exported in the form of pulpwood to Norway, Sweden, Italy and Portugal. It is also used for local consumption, mainly

73

for poles and for charcoal production. The organization is affiliated to the Ministry of Forestry.

Beside state organizations, logging is also carried out by private foreign firms, mixed Congolese–foreign companies and private Congolese entrepreneurs.

In agreement with the Forest Law No. 4/1974 and the Decree to Forest Law No. 84/910, management units were established in 1984 in the Congolese forest massifs for the regulation and organization of logging. These management units may look after an area of 0.5 million ha or more. They are subdivided into logging blocks which are allocated by the Ministry of Forestry in the form of contracts or permits to state-owned or other logging organizations.

Keeping in line with rules and regulations, there are so-called industrial and exploitation contracts in Congo which are issued to logging companies for a certain area, e.g. 60,000 ha and for a certain time period, e.g. 15 years. For smaller Congolese enterpreneurs, permits are issued for logging certain tree species to a determined number of trees, usually from 100 to 400, for a period of one year. Special permits are issued for the exploitation of a certain amount of timber or of other forest products for consumption by the rural population and are issued for the period of one month.

The forest legislation of the Republic of Congo regulates the main forestry activities. Alas, no solutions are given for the needs of the local population inhabiting forest regions, for its social problems, for land use and soil conservation, and for problems of the natural and living environment. Although there are provisions for the development of forestry, no realistic agricultural programme is in force for the forest regions of the country.

4.2.3 Results of the forest inventory
    in the Sibiti-Zanaga region

The Sibiti-Zanaga region is situated in the area of Lékoumou, in the tropical moist climate zone. Average annual precipitation in this region is around 1,500 mm with two maxima of the rainy season in October and November, and again in March and April, and with one dry period from June to the middle of September. The relative air humidity is high throughout the year, not declining below 75%. The mean annual temperature is 23–24° C. The daily temperature during the year is without major fluctuations.

Forest inventory in the Sibiti-Zanaga region was made during 1971–1972 under FAO project by a group of French and Czechoslovak foresters, in colaboration with Congolese forest authorities.

All forests occurring in the Sibiti-Zanaga region may be classified as belonging to the formation of the moist evergreen or moist semi-deciduous tropical forest. Both forest formations occur mostly on solid soil, seldom on waterlogged or periodically inundated land.

Fig. 39. Moist evergreen forest with the trees of *Aucoumea klaineana* (on the right) and *Klainedoxa gabonensis* (on the left). Sibiti-Zanaga region, Congo.

The formation of the moist evergreen forest is characterized by luxuriant growth and diversified vegetation, by a large number of tree species and trees of different ages which keep their foliage throughout the year. The tallest trees of the uppermost storey attain a height of 50–60 m and a diameter of 150 cm and more. Some tree species grow only to a diameter of 20–30 cm and a height of 7–10 m. The inferior storey consists of various herbs, grasses and shrubs. Lianas of different shape, epiphytes and saprophytic plants frequently occur in the forest. Some trees form big, 3–5 m high buttresses, these include *Klainedoxa gabonensis*, *Aucoumea klaineana* (Fig. 39), *Piptadeniastrum africanum* and *Nauclea diderrichii*.

Moist semi-deciduous tropical forests form a transition belt between evergreen and moist deciduous tropical forests. They spread over different sites throughout the entire inventoried region. They, too, comprise several storeys. In contradistinction to evergreen moist tropical forests certain tree species in semi-deciduous tropical forests shed their foliage in the dry season. As, for example, do *Terminalia superba, Chlorophora excelsa, Pterocarpus soyauxii, Ceiba pentandra* and other tree species.

The following additional forest types were identified in the inventoried region during field-work:

Swampy forests occur on waterlogged soil and in the vicinity of rivers; *Mitragyna ciliata, Guarea cedrata, Nauclea diderrichii*, species of the *Pandanus* genus and other tree species, are typical of such forests.

Periodically inundated forests occur in valley and near to rivers. The soil is periodically inundated in the rainy season, usually for a period of a few weeks. The following tree species are typical for periodically inundated forests in the Sibiti-Zanaga region: *Guibourtia demeusei, Oxystigma oxyphyllum*, species of the *Uapaca* genus, *Gilbertiodendron dewevrei*, etc. Palm trees of the *Raphia* genus are frequently represented in the under-storey.

Forest stands with umbrella crowns are a secondary forest consisting mainly of fast growing, light demanding tree species frequently intertwined with various lianas. Here, *Musanga cecropioides,* a tree with typical umbrella crowns is the main species. Among commercial species *Fagara heitzii* occurs frequently in this forest type.

Young, naturally regenerated secondary forest stands form in places where previously native peoples cultivated agricultural crops for a few years. Commercial tree species, such as *Aucoumea klaineana, Fagara heitzii, Baillonella toxisperma*, occur in such forests, too. But mostly species of the *Macaranga, Vernonia* and *Musanga* genera are represented. The oil palm, *Elaeis guineensis*, is also to be found.

The area of the Sibiti-Zanaga inventoried region was 976,650 ha, which included 828,600 ha of productive forests, or 85% of the inventoried region. The area of degraded forests and deforested areas was 117,200 ha, i.e. 12% of the region. Because of the steepness of the terrain, 30,850 ha of forest was classified as a protection forest. Originally it was intended to divide the region into three categories of inventory blocks according to size, namely into large blocks with an area of about 60,000 ha, medium sized blocks with a surface of about 25,000 ha, and small blocks with a surface of about 5,000 ha.

After thorough consideration with regard to the best use of available funds, and after an aerial reconnaissance of the terrain, the whole region was divided into 12 large blocks with an average area of 57,479 ha (73% of the inventoried region), 7 medium sized blocks with an average area of 37,314 ha (25%) and 3 small inventory blocks with an average area of 8,567 ha (2% of the inventoried region). The inventory blocks were intended in the future to assume the function of logging blocks – after awarding a contract to the logging company in question. The area of all inventory blocks is given in Table 14.

Assuming that the timber stock in each block will be determined within an accuracy of ±20% at a 5% level of significance, the intensity of extraction (sampling) was as follows (see Table 14).

Table 14. The area of inventory blocks. Sibiti-Zanaga region

Number of blocks	Productive forest ha	Deforested area ha	Protection forest ha	Total area ha
Large blocks – intensity of sampling: 0.22%				
1	65,700	8,100	–	73,800
2	52,450	3,250	–	55,700
3	67,000	9,400	–	76,400
5	43,100	3,950	–	47,050
11	50,100	3,000	28,650	81,750
12	65,900	16,700	–	82,600
16	56,900	1,300	–	58,200
18	38,800	–	–	38,800
19	33,400	4,450	–	37,850
20	56,550	1,300	–	57,850
21	33,900	7,150	–	41,050
22	38,700	–	–	38,700
Total	602,500	58,600	28,650	689,750
Medium blocks – intensity of sampling: 0.44%				
4	27,150	15,150	–	42,300
7	25,700	13,550	–	39,250
8	29,500	6,100	–	35,600
10	25,100	4,300	–	29,400
13	39,350	7,650	–	47,000
15	36,650	4,150	–	40,800
17	26,850	–	–	26,850
Total	210,300	50,900	–	261,200
Small blocks – intensity of sampling: 2.50%				
6	5,300	550	–	5,850
9	4,900	4,800	2 200	11,900
14	5,600	2,350	–	7,950
Total	15,800	7,700	2 200	25,700
Totals	828,600	117,200	30 850	976,650
%	84,8	12,0	3,2	100

The inventory was made statistically, by systematic sampling in line plots oriented at 370° to the magnetic pole. The distance between inventory lines was 6 km in large blocks, 3 km in medium sized blocks and 250 m in small blocks. The inventoried plots on the inventory lines were of a rectangular shape having the dimensions of 25 × 200 m (= 0.50 ha) in large and medium sized blocks and 25 × 50 m (= 0.125 ha) in small blocks. The distance between the inventoried plots on the lines was 200 m in large and medium sized blocks and 150 m in small blocks (Fig. 40). A total of 2,737 km of inventory lines was marked and cleared. The inventory was made on 4,810 inventoried plots each having an area of 0.50 ha either made up into large or medium sized blocks and in total covering 2,405 ha. And on 3,055 inventoried plots each with an area of 0.125 ha in small inventory blocks in total covering 382 ha. The total area of inventoried plots was 2,787 ha. The tree species recorded by the inventory were classified according to their economic importance into the following four categories:

1. Main tree species of primary importance.
2. Main tree species of lesser importance.
3. Secondary tree species (lesser-known species).
4. Remaining tree species.

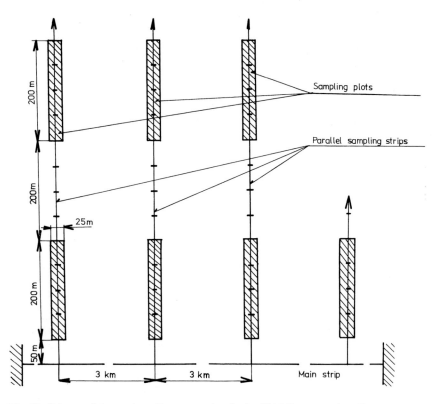

Fig. 40. Scheme of the tracing of inventory plots in the Sibiti-Zanaga region, Congo.

The main tree species of primary importance included:
   Aucoumea kaineana PIERRE – okoumé
   Lovoa trichilioides HARMS – dibetou
   Baillonella toxisperma PIERRE – moabi

Turraeanthus africana PELLEGR. – avodiré
Entandrophragma angolense C. DC. – tiama
E. candollei HARMS – kosipo
E. cylindricum SPRAGUE – sapele (sapelli)
E. utile SPRAGUE – utile (sipo)
Dumoria africana A. CHEV. – douka
Chlorophora excelsa BENTH. et HOOK. f. – iroko
Afzelia bipindensis HARMS – afzelia (doussié)
Terminalia superba ENGL. et DIELS – limba
Millettia laurentii DE WILD. – wengé
Guarea cedrata PELLEGR. – guarea (bossé)
Testulea gabonensis PELLEGR. – izombé
Swartzia fistuloides HARMS – dina (pau rosa)
Microberlinia brazzavillensis A. CHEV. – zingana
The main species of lesser importance include:
Fagara heitzii AUBR. et PELLEGR. – olon
Oxystigma oxyphyllum J. LÉONARD – tchitola
Pycnanthus angolensis WARB. – ilomba
Staudtia stipitata WARB. – niové
Pterocarpus soyauxii TAUB. – African padauk
Distemonanthus benthamianus BAILL. – ayan (movingui)
Canarium schweinfurthii ENGL. – African canarium (aiélé)
Daniellia ogea ROLFE – faro
Nauclea diderrichii MERR. – bilinga
Dacryodes buettneri H. J. LAM. – ozigo
Mitragyna ciliata AUBR. et PELLEGR. – abura (bahia)
Gilbertiodendron dewevrei J. LÉONARD – limbali
The lesser-known tree species include:
Dacryodes pubescens H. J. LAM. – safoukala
Alstonia congensis ENGL. – emien
Coelocaryon preusii WARB. – ekoune
Scyphocephalium ochocoa WARB. – soro (sogho)
Piptadeniastrum africanum BREMAN – dabema
Berlinia bracteosa BENTH. – ebiara
Celtis brieyi DE WILD. – diania
Gambeya africana PIERRE = *Chrysophyllum africanum* C. DC. – longhi
Anopyxis klaineana ENGL. – bodioa
Ricinodendron africanum MUELL. ARG. – essessang
Amphimas ferrugineus PIERRE ex PELLEGR. – lati
Those species which may be exported in the form of logs, sawn timber or veneer, were classified as main commercial tree species of primary importance. Among them, the most important tree species include okoumé and the so-called red timbers, i.e. species of the *Entandrophragma*, *Guarea* and *Lovoa* genera used mainly

for decorative veneer, and also moabi, avodiré, wengé as well as other species.

Tree species of less important economic use are destined at present mostly for domestic processing. Olon, ilomba, niové, bilinga, diambi, tchitola, aiélé and other species occur almost throughout the entire country.

At present there is a modest market demand for secondary tree species such as emien, safoukala, ekoune, dabema and ebiara.

The other 360 tree species in the inventoried region are not exploited at present because of lack of market. By today's utilization standards, most of the trees in the humid tropics are, from an industrial-materials standpoint, clearly weeds (Bethel, 1984). The most common such species recorded in the inventory region include: *Uapaca guineensis* – rikio, *Klainedoxa gabonensis* – eveuss, *Pentaclethra macrophylla* – mubala, *Combretodendron africanum* – essia (wulo), *Erythrophleum guineense* and *E. suaveolens* – tali, *Gilbertiodendron dewevrei* – limbali, *Monopetalanthus pellegrini* – mayo (andoung) and *Strombosia gossweileri* – ekumba-apio.

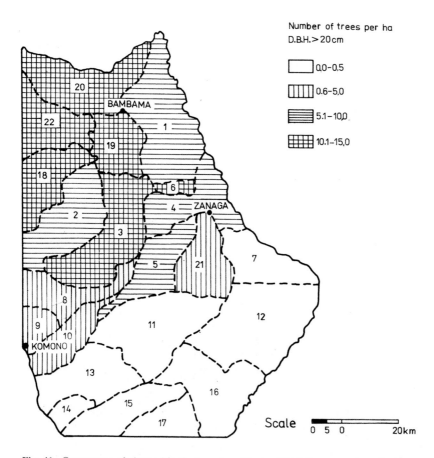

Fig. 41. Occurrence of okoumé in the inventory blocks. Sibiti-Zanaga region, Congo.

The callipered trees were classified at 10 cm diameter intervals, starting with a diameter breast height of 20 cm for commercial and lesser-known tree species and a diameter breast height of 60 cm for other tree species. Tree diameters were measured either indirectly with a special calliper at an inaccessible height above the buttresses, or with a tape around the girth of the stem at breast height.

Okoumé occurs frequently in the northern parts of the inventoried region, in the south it is absent (Fig. 41). Mahoganies are less frequent and they grow sporadically. Limba is a rare species occurring in the southeast of the inventory region. Agba was not recorded in any block.

Among the main tree species of primary importance only dibetou, and among the main tree species of secondary importance ilomba, bahia and olon were recorded in each inventoried block. From among the lesser-known tree species dabema and ekoune occurred throughout the entire inventoried region. From among other tree species recorded from a diameter of 60 cm upwards, 70–130 different tree species occurred in the various inventory blocks, only rikio being represented in each block.

Some tree species grow to a diameter of 100 cm, including ekoune, bossé, wengé; others to a diameter of 120 cm, such as padouk, dibetou, limba, doussié, still other species attain a diameter of 150 cm or more, for example okoumé, moabi, utile, kosipo, tiama, dabema and limbali.

Considering a single tree species or its community, the distribution of trees in the forest by their diameter breast height, is an important indicator of the diameter structure of the tropical forest and of potential timber assortments derived from it. Bearing this in mind, the author has expressed this regularity using the well-tried mathematical formula of the beta function.

The beta function was used for the theoretical expression of the diameter distribution of trees of certain main and lesser-known species (Table 15, Fig. 42). It was also used in aggregation for the main and secondary species in the entire inventoried region. Almost all species showed a declining tendency of the diameter distribution of trees. Although the frequency distribution of trees of certain tree species by their diameter did not correspond to the beta distribution with regard to the resulting value of the χ^2 test, the beta function can be considered to be a suitable mathematical formula for the expression of the diameter distribution of trees in natural closed tropical forests in the inventoried region in Congo.

In the whole region the productive and degraded forests were recorded separately. Data given in Table 16 show a much larger number of trees of all categories of species in the productive forest as compared to the degraded one where the negative consequences of shifting cultivation are apparent (Fig. 43). Mean numbers of trees per hectare are one of the important survey indicators in natural high tropical forests. The various tree species and their categories were classified into diameter classes with an interval of 20 cm in the whole inventoried region. For all tree species the same felling diameter and exploitation limits, namely 60 cm were imposed.

In the productive forests of the inventoried region okoumé represents, from

Table 15. Theoretical expression of the beta-diameter distribution of trees of some timber species. Sibiti-Zanaga region

Midpoint value of D.B.H. class	Species				
	Okoumé	Ilomba	Bilinga	Avodiré	Padouk
cm	Number of trees per 1,000 ha				
25	3,027	3,288	1,158	703	196
35	2,284	2,377	677	478	186
45	1,768	1,409	377	308	159
55	1,357	747	194	188	129
65	1,022	358	87	108	101
75	750	156	36	60	73
85	530	61	9	30	49
95	358	19	1	14	29
105	220	4	–	6	15
115	122	–	–	3	2
125	54	–	–	1	–
135	17	–	–	–	–
145	2	–	–	–	–
α	−0.07506	0.23099	−0.04285	0.04560	0.11989
β	2.37754	6.49915	3.97955	6.17066	1.71688
Const.	0.0036	0.0000	0.0000	0.0000	0.0067

among all main tree species, one half of all exploitable trees (D. B. H. ≥ 60 cm). If trees with a diameter exceeding 20 cm are taken into consideration, okoumé represents 33% of the main species represented.

From among all species with a diameter of 60 cm and more, the category of unused tree species included 53% of the trees.

In the category of main tree species, in addition to okoumé, the following species were most represented in the productive forest: ilomba with 3.45 trees per hectare, olon with 1.36 trees per hectare, bahia with 1.34 trees per hectare, bilinga with 1.05 trees per hectare, niové with 1.02 trees per hectare, and avodiré with 0.96 trees per hectare. From among secondary tree species, ekoune with 11.14 trees per hectare, sogho with 1.73 trees per hectare, and dabema with 1.48 trees per hectare are the most frequent. Okoumé was the most common commercial tree species also present in the degraded forests of the inventoried region, but only with 1.7 trees per hectare had a D.B.H. exceeding 20 cm.

In order to determine the timber stock in the Sibiti-Zanaga inventory region it was necessary to set up volume tariffs for the main species and for groups of certain tree species. In the process of setting up volume tariff for okoumé, dendrometrical values were determined for 88 sample trees freshly felled in the Mosendjo logging region in the vicinity of the inventoried region. According to usage, the log volume

Fig. 42. Diameter distribution of trees for the main tree species according to the beta function. Sibiti-Zanaga inventory region, Congo.

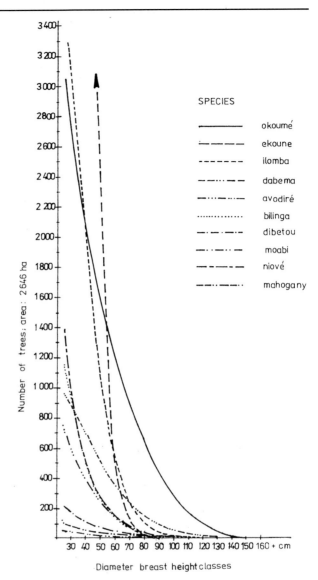

SPECIES

——————	okoumé
— — — —	ekoune
- - - - - -	ilomba
— · — · —	dabema
— ··· — ··· —	avodiré
··············	bilinga
— · — · — ·	dibetou
— ·· — ·· —	moabi
— — · — —	niové
— ··· —— ··	mahogany

Number of trees; area: 2 646 ha

Diameter breast height classes

was measured from the buttresses to the first branch of the tree crown. The volume of the felled sample trees was computed using Smalian's formula by sections (Korf et al., 1972)

$$V = l' \left(\frac{A_0 + A_n}{2} \right) + (A_{1/3} + A_{2/3}),$$

where V is the volume of the log, l' the length of the log section, A_0, $A_{1/3}$, $A_{2/3}$, A_n are cross-sectional areas (Fig. 44).

83

Table 16. Mean numbers of trees per hectare in the productive and degraded forest of the inventoried Sibiti-Zanaga region

Species	Diameter classes in cm				D.B.H. > 60 cm	Total
	20–40	40–60	60–80	80+		
Productive forest						
Okoumé	2.60	1.40	0.95	0.57	1.52	5.52
Other main species	7.50	3.06	1.17	0.40	1.57	11.13
Secondary species	11.95	2.53	1.27	0.44	1.71	16.19
Together	22.05	6.99	3.39	1.41	4.80	33.84
Other species	–	–	4.26	0.95	5.21	5.21
Total	22.05	6.99	7.65	2.36	10.01	39.05
Degraded forest						
Okoumé	0.96	0.41	0.17	0.17	0.34	1.71
Other main species	3.95	1.35	0.57	0.11	0.68	5.98
Secondary species	6.15	0.97	0.55	0.14	0.69	7.81
Together	11.06	2.73	1.29	0.42	1.71	15.50
Other species	–	–	1.63	0.37	2.00	2.00
Total	11.06	2.73	2.92	0.79	3.71	17.50

Field measurements included the girth of the stem of the sample trees measured at the base of the stem, at one third and two thirds of the stem length, and at the top of the stem. The diameter breast height of the felled sample trees of okoumé varied between 70 and 150 cm ($\bar{d}$ = 101.9 cm), the length of the sample trees between 8 and 28 m ($\bar{l}$ = 18.2 m), and the computed volume of the various stems between 1.7 and 21.5 m³.

The homogeneity or heterogeneity of the empiric material were determined by the correlation dependence between the measured characters of the stems, i.e. by the diameter breast height (d_i) and length (l_j) of the stems (see correlation Table 17).

The computed value of the correlation coefficient r = −0.115 indicates that the correlation dependence between the diameter and the length of the felled sample trees is very small (Freese, 1962).

For setting up volume tariffs for okoumé a very heterogeneous empirical material was used (see Table 18) which is indicative of the great growth diversity of trees in the tropical moist evergreen forest. In setting up volume tariff for okoumé the author tried several mathematical functions. For testing their suitability Gauss' criterion was used. The best suited function was

$$V = a\, x^b,$$

with weighted means, where V is the log volume, x the diameter (D.B.H.), a, b are constants.

After substitution the following equation was obtained:
$$V = 0.000507\, x^{1.98864}.$$

Fig. 43. Shifting cultivation in a tropical forest. Region of Chaillu, Congo.

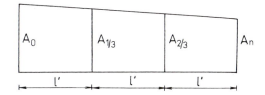

Fig. 44. Scheme of the measurement of sample trees for calculating their volume.

85

Table 17. Correlation between breast height diameter (d_i in cm) and length (l_j in m) of okoumé felled stems. Mossendjo, Congo

d_i \ l_j	8	10	12	14	16	18	20	22	24	26	28	Σn_i	$\bar{l}_j$	$n_i d_i$	$n_i d_i^2$	$n_{ij}d_i l_j$
70			1									1	12.0	70	4,900	840
80	1		1	1	3	4	2	1	1	1		15	17.7	1,200	96,000	21,200
90			1	2	3	3	4		4	1		18	19.1	1,620	145,800	30,960
100			1	1	4	2	3	3	3	1	1	19	19.9	1,900	190,000	37,800
110		1	2	1	1	6	3		2			16	17.5	1,760	193,600	30,800
120		1			3	2	2	1				9	17.3	1,080	129,600	18,720
130		1	1	1	2	1		1				7	15.4	910	118,300	14,040
140					1		1					2	18.0	280	39,200	5,040
150						1						1	18.0	150	22,500	2,700
Σn_j	1	3	7	6	17	19	15	6	10	3	1	88	–	8,970	939,900	162,180
$\bar{d}_i$	80.0	120.0	98.6	116.7	104.7	103.7	102.0	105.0	96.0	90.0	100.0					
$n_j l_j$	8	30	84	84	272	342	300	132	240	78	28	1,598				
$n_j l_j^2$	64	300	1,008	1,176	4,352	6,156	6,000	2,904	5,760	2,028	784	30,532				
$n_{ij} l_j d_i$	640	3,600	8,280	8,400	28,480	35,460	30,600	13,860	23,040	7,020	2,800	162,180				

$$\bar{d} = \frac{1}{n}\Sigma n_i d_i = \frac{8970}{88} = 101.93,$$

$$\bar{l} = \frac{1}{n}\Sigma n_j l_j = \frac{1598}{88} = 18.16,$$

$$s_d = \sqrt{\frac{\Sigma n_i d_i^2}{n} - \bar{d}^2} = \sqrt{\frac{939900}{88} - 101.93^2} = 17.057,$$

$$s_l = \sqrt{\frac{\Sigma n_j l_j^2}{n} - \bar{l}^2} = \sqrt{\frac{30532}{88} - 18.16^2} = 4.144,$$

$$r = \frac{\Sigma n_{ij} d_i l_j - n\,\bar{d}\,\bar{l}}{n\,s_d s_l} = \frac{162{,}180 - 88 \times 101.93 \times 18.16}{88 \times 17.053 \times 4.144} = -0.115.$$

Table 18. Volume tariff for okoumé

Midpoint value of D.B.H. class	Felled sample trees			Calculated volume of the stem
	Number of	Volume of stems	Volume of main stem	
cm	stems	m³		
50				1.21
55				1.47
60				1.74
65				2.04
70	1	1.69	1.69	2.37
75	3	8.58	2.86	2.72
80	6	18.96	3.16	3.08
85	10	40.00	4.00	3.48
90	9	38.34	4.26	3.90
95	7	34.16	4.88	4.35
100	14	64.24	4.59	4.81
105	9	50.13	5.57	5.30
110	6	36.48	6.08	5.82
115	7	44.31	6.33	6.35
120	3	19.35	6.45	6.92
125	3	15.75	5.25	7.50
130	3	23.22	7.74	8.11
135	4	33.16	8.29	8.74
140	1	9.07	9.07	9.40
145	2	21.84	10.92	10.07
150				10.78
155				11.50
160				12.25
165				13.03
170				13.82
175				14.64
180				15.49
Total	88	459.3	5.22	–

For testing, use was made of: (a) Gauss' criterion = 0.07693, (b) correlation index $i = 0.98692$.

The calculations were made using the "Tesla 200" computer at the University College of Forestry and Wood Technology in Zvolen. Empirical material for the construction of volume tariffs for the main tree species and groups of species was obtained by measuring inaccessible diameters and heights of standing trees using Bitterlich's relascope with a wide scale. The following uniform formula was used for the construction of volume tariffs:

$$V = a + b\, D^2,$$

where V is the stem volume, D the D.B.H. of trees, a, b are constants.

Constants computed for some tree species are listed below:

dibetou	$a = -0.14062$	$b = 9.84054$
moabi	$a = -0.30499$	$b = 9.60409$
avodiré	$a = -0.07853$	$b = 7.82389$
tchitola	$a = -0.28790$	$b = 11.05470$

The calculations were made by computers of the Centre Technique Forestier Tropical (C.T.F.T.) in Nogent sur Marne, France.

Figure 45 shows (a) the diameter distribution of all trees of okoumé recorded on sample plots in the Sibiti-Zanaga inventoried region over an area of 2,240 ha and (b) the relating timber stock computed according to the mentioned volume tariffs.

The average growing stock per hectare computed for okoumé and other

AUCOUMEA KLAINEANA

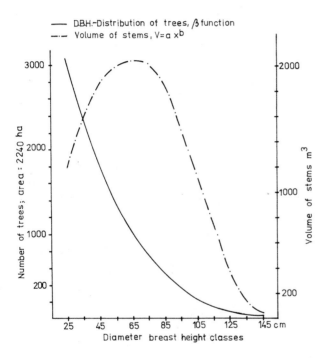

Fig. 45. Diameter distribution of trees for okoumé according to the beta-function and their timber stock. Sibiti-Zanaga region, Congo.

categories of tree species in both the productive and degraded forest in the inventoried region is given in Table 19. The following conclusions may be drawn from these data: the growing stock of exploitable trees of okoumé in the productive forest is 8 m³ per hectare. The other main tree species of primary and secondary importance have likewise a timber stock of 8 m³ per hectare. From among commercial tree species utile, iroko, limba and sapele are of limited occurrence. Secondary tree species with an average timber stock of 10 m³ per hectare are represented favourably. Much more represented is, of course, the group of unused tree species with

Table 19. The average growing stock in m^3 ha^{-1} for okoumé and other categories of tree species. Sibiti-Zanaga inventoried region

Species	Diameter classes in cm				D.B.H. ≥ 60 cm	Total
	20–40	40–60	60–80	80+		
Productive forest						
Okoumé	1.59	2.61	3.54	4.52	8.06	12.26
Other main species	5.76	6.19	4.69	3.57	8.26	20.21
Secondary species	7.24	5.03	5.88	4.02	9.90	22.17
Together	14.59	13.83	14.11	12.11	26.22	54.64
Other species	–	–	21.14	9.39	30.53	30.53
Total	14.59	13.83	35.25	21.50	56.75	85.17
Degraded forest						
Okoumé	0.59	0.77	0.65	1.43	2.08	3.44
Other main species	2.81	2.66	2.33	1.00	3.33	8.90
Secondary species	3.51	1.90	2.46	1.25	3.71	9.12
Together	7.01	5.33	5.44	3.68	9.12	21.46
Other species	–	–	8.14	3.51	11.65	11.65
Total	7.01	5.33	13.58	7.19	20.77	33.11

a timber stock of 30 m^3 per hectare. The degraded forest is of smaller productivity, containing a lower supply of exploitable timber at around 21 m^3 per hectare including 9 m^3 per hectare of commercial and lesser-known species.

In the productive forest of the entire inventoried region the following tree species have the largest share in the production of mature timber (D.B.H. ≥ 60 cm):

No.	Species	% of total production
1	okoumé	14.2
2	rikio	8.2
3	eveuss	4.2
4	sogho	3.8
5	essessang	2.9
6	limbali	2.7
7	diania	2.6
8	eba	2.5
9	dabema	2.1
10	essia	2.0
Total		45.2

The timber stock of the 10 tree species mentioned represents 45% of all the timber production exceeding the diameter of 60 cm in the inventoried region. The other 390 tree species occurring in this region and growing to a diameter of 60 cm and more, make up only 55% of the total timber production. From among the ten tree species mentioned, only okoumé may be envisaged for export. There are no consumers for the other 9 tree species.

The total production of the timber stock for trees with a diameter exceeding 60 cm in the productive forest of all inventoried blocks by tree species categories was as follows:

okoumé	6,567,000 m^3 =	14.2%,
other main tree species 1	1,881,200 m^3 =	4.1%,
other main tree species 2	4,857,700 m^3 =	10.4%,
secondary tree species	8,074,400 m^3 =	17.5%,
other tree species	24,896,800 m^3 =	53.8%,
Total	46,277,100 m^3 =	100,0%.

Converted to one hectare, this is a timber stock of 56 m^3.

4.2.4 Forest management in the Sibiti-Zanaga region

Forest management of high tropical forests based on the principle of sustained yield and assured reproduction of the tropical forest is of immense importance for the dependable integral development of the countryside. This has got to be done in harmony with the development of forestry, forest industries, agriculture, the observation of ecological requirements and the conservation of forest resources in any developing country.

Forest management may be put into practice only when a sufficient and qualified staff of forest personnel is available to carry out the necessary tasks in forest utilization. In addition to logging operations they also need to take charge of silvicultural work, forest regeneration and tending, and the associated problems of the rural population (Dawkins, 1958). Because in the Republic of Congo in the early 1970s, there was a lack of qualified foresters, only elementary steps could be taken in forest management. This could be later improved and enlarged with the development of forestry.

For the purpose of forest management, the whole inventoried region was divided into: (a) productive forests, (b) protection forests, and (c) degraded forests.

(a) Productive forests cover an area of 828,600 ha, i.e. 85% of the total surface of the inventoried region.

(b) Protection forests are situated in regions with steep slopes. Logging is not recommended. The main function of the forest is soil conservation. Protection forests cover an area of 30,850 ha in the inventoried blocks 9 and 11, i.e. 3% of the total surface.

(c) Degraded forests including agricultural plantations in the forest, take up an area of 117,200 ha, i.e. 12% of the total area of the inventoried region.

Forest management concerns the productive forest. It does not solve detailed problems of the protection forests, nor problems of the devastated forest or of agricultural plantations.

Okoumé being, from the point of view of the national economy of the Republic of Congo, the most important tree species, two management units were marked off in the inventoried region, namely:

(a) the management unit containing okoumé covering an area of 487,650 ha, i.e. 59% of the area of productive forests,

(b) the management unit with various other tree species covering an area of 340,950 ha, i.e. 41% of the area of productive forests.

As a criterion for classifying the inventory blocks into one or the other management unit, the number of okoumé trees with a diameter exceeding 60 cm per hectare was used (Borota, 1973).

The mean number of trees and their growing stock per hectare for okoumé and the various categories of tree species for the entire management unit containing okoumé are given in Table 20. Among other tree species which occur most frequently with okoumé, there is sogho in the northeastern part and movingui in the northwestern part of the inventoried region. Sogho (soro) is a secondary tree species growing to a diameter of 150 cm. It is so far of no commercial importance. Movingui belongs to the category of main tree species of secondary importance. Its timber is used in furniture making, construction, carpentry and for the production of veneer.

Table 20. The mean number of trees and their growing stock per hectare for okoumé and the various categories of tree species. Management unit with okoumé

Species	Number of trees per ha		Growing stock in m³ per ha	
	D.B.H. = 20–60 cm	D.B.H. > 60 cm	D.B.H. = 20–60 cm	D.B.H. > 60 cm
Okoumé	6.75	2.55	7.06	13.33
Other main species 1	1.47	0.27	1.51	1.61
Other main species 2	9.59	1.22	10.99	5.58
Secondary species	18.27	1.85	15.75	11.45
Other species	–	4.37	–	16.74

According to inventory results, the total production of mature trees, i.e. those with a diameter breast height ≥ 60 cm, amounts to 23,750,000 m³ (in an area of 487,650 ha) in the management unit containing okoumé. From among the timber resources of mature trees the share of okoumé is 27%, the share of other main tree species of primary importance is 3%, the share of main tree species of secondary importance is 11.5%, the share of secondary tree species is 23.5% and the share of other, so far unmarketable tree species is 35% of the timber stock.

Table 21. Number of trees of okoumé with a diameter breast height above 20 cm in 160 inventoried sample plots. Selected territory of 2×4 km of the inventory block No. 6, Sibiti-Zanaga region

Sample plots	Inventory lines															
	a	b	c	d	e	f	g	h	i	j	k	l	m	n	o	p
I	4	1	—	6	1	2	4	—	5	—	—	—	2	3	1	—
II	—	2	3	1	—	1	—	2	1	1	3	—	—	2	1	1
III	1	2	4	—	—	7	—	1	—	—	1	2	1	1	2	2
IV	4	4	4	—	1	1	—	—	—	3	2	—	5	3	1	14
V	—	1	1	—	—	1	1	—	1	3	1	—	5	1	4	3
VI	—	—	—	2	3	—	—	3	—	—	—	4	1	2	1	1
VII	—	1	—	1	1	—	—	1	1	1	1	2	3	—	1	1
VIII	—	2	8	1	2	—	—	1	1	4	1	1	2	—	—	—
IX	3	—	5	3	1	—	3	1	—	3	3	—	—	—	2	—
X	1	1	2	5	2	3	1	2	1	2	1	—	—	4	—	2
Total	13	14	27	19	11	15	9	10	10	17	13	9	19	16	13	24
Number of trees per ha	10,4	11.2	21.6	15.2	8.8	12.0	7.2	8.0	8.0	13.6	10.4	7.2	15.2	12.8	10.4	19.2

Block No. 22 is the inventory block with the largest amount of okoumé. Here, the growing okoumé stock was 19 m³ per hectare (for trees whose diameter exceeded 60 cm).

As a practical illustration, the number of trees of okoumé with a diameter above 20 cm is given in Table 21 for 160 inventoried sample plots in a selected territory of 2 × 4 km of the inventory block No. 6, i.e. over an area of 800 ha where okoumé is frequent.

In the block No. 6 the inventory covered an area of 2.5% of the block. The width between the inventory lines was 250 m and the distance between the plots which had an area of 0.125 ha (50 × 25 m) was 150 m.

Data given in Table 21 show that okoumé occurred on the sample plots irregularly. In relation to the area of 1 ha, there were, on average, 12 trees of okoumé with a diameter breast height more than 20 cm and 4 trees of okoumé with a diameter greater than 60 cm. From among 160 inventory plots there were: on 55 plots 34% okoumé trees with a diameter over 20 cm did not occur; on 109 plots, i.e. 68% of all inventory plots, there were no mature trees of okoumé with a D.B.H. greater than 60 cm. The largest number of recorded okoumé trees on a plot (= 0.125 ha) was 14 – all of them with a D.B.H. of 20–60 cm, and on an other sample plot there were 4 mature trees. This gives evidence of a great irregularity in the representation of okoumé, even in forests where it occurs frequently. This fact should be borne in mind when considering informative random surveys regarding the occurrence of okoumé in a certain part of the forest.

The management unit with various tree species covered an area of 340,950 ha of productive forests. Average numbers of trees and their growing stock per hectare for okoumé, and for the various categories of tree species in the management unit with different tree species are given in Table 22.

Table 22. The mean number of trees and their growing stock per hectare for okoumé and the various categories of tree species. Management unit with different tree species

Species	Number of trees per ha		Growing stock in m³ per ha	
	D.B.H. = 20–60 cm	D.B.H. > 60 cm	D.B.H. = 20–60 cm	D.B.H. > 60 cm
Okoumé	0.69	0.04	0.10	0.20
Other main species 1	1.34	0.51	1.62	3.23
Other main species 2	8.51	1.19	9.56	7.03
Secondary species	9.11	1.49	7.34	8.62
Other species	–	6.54	–	38.43

Inventory blocks where okoumé occurs in a negligible amount or is absent altogether, are not notable for other primary commercial tree species either. The major representation and the majority of timber resources belonged to the so-called other tree species without economic importance.

From among commercial tree species occurring sporadically in this management unit the African mahoganies should be mentioned in particular. Among main tree species of secondary economic importance ilomba, limbali, bahia, niové, aielé, padouk, and tchitola occur most frequently. And among secondary tree species mention should be made of ekoune, dabema, safoukala, ebiara, diania and essessang.

The total production of mature trees in the management unit with various tree species was, according to inventory data, about 19,600,000 m^3; of this 67% belongs to species which have no economic importance, 15% to secondary tree species, 12% to main species of secondary importance, and only 6% to main tree species of primary importance.

In the region of evergreen moist tropical forests it is very difficult to determine the annual diameter or volume increment of trees for the various species. In the Sibiti-Zanaga inventory region no long-term observational growth data or measurements of the increment of commercial tree species were available.

The average D.B.H. increment in natural tropical forests varies with many factors, but is seldom greater than 1 cm per year and is often less (Schmidt, 1987). According to observations made in neighbouring Gabon the mean annual diameter increment for okoumé varies from 0.8 to 1.0 cm. From certain hypotheses derived from inventory data the logging cycle in both management units, was determined for 20 years. The growing stock of all commercial tree species with a diameter exceeding 60 cm was taken into consideration. In the management unit containing okoumé this resulted in about 6,500,000 m^3 in the first logging cycle. In the second logging cycle a growing stock of about 3,250,000 m^3 was expected which represents about 50% of the timber volume determined for the first logging cycle.

Both management units were divided into two felling series. In both of them, fellings should have taken 10 years. The scheme of fellings is shown in Fig. 46.

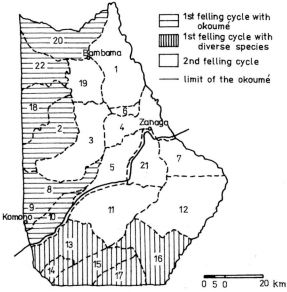

1st felling cycle with okoumé
1st felling cycle with diverse species
2nd felling cycle
limit of the okoumé

0 5 0   20 km

Fig. 46. Pattern of logging cycles in the Sibiti-Zanaga inventory region, Congo.

The area of productive forests in the management unit containing okoumé, where logging was envisaged in the first decennium (blocks Nos 2, 8, 9, 10, 18, 20, 22) was 246,000 ha. This means that annual area cut in this section was 24,600 ha. Taking into account inventory data regarding the mean number of trees per hectare, and with a view to the possibility of timber utilization in Congo, this annual cutting area corresponds to a growing stock of about 725,000 m³. This annual cut includes:

okoumé	327,000 m³,
other main tree species	176,000 m³,
secondary tree species	209,000 m³,
other tree species	13,000 m³.

In the following decennium fellings will take place in blocks Nos 1, 3, 4, 5, 6 and 19, their area being 241,650 ha. The annual allowable cut of 24,165 ha corresponds to an annual allowable cut of:

okoumé	322,000 m³,
other main tree species	154,000 m³,
secondary tree species	324,000 m³,
other tree species	10,000 m³.

The average recorded felling of logs had to be 33 m³ per hectare. For each inventoried or logging block more detailed forecasts were made concerning the logging possibilities of certain tree species. Recommendations were made for the allotment of entire inventory blocks to logging companies so as to convert them into logging blocks.

A similar procedure was envisaged for the management unit containing various tree species. For the first decennium, block Nos 13 to 17 with an area of 165,350 ha were earmarked for logging operations. The total timber volume to be felled annually, corresponding to the annual cutting area of 16,535 ha, was 340,000 m³. This included annually:

main tree species including okoumé	168,000 m³,
secondary tree species	160,000 m³,
other tree species	12,000 m³.

In other words, about 20 m³ of timber per hectare could be extracted annually.

In the following decennium block Nos 7, 11, 12, and 21 (with a total area of 175,600 ha of productive forest) were earmarked for exploitation. Expected annual fellings amounted to about 304,000 m³ of logs over an area of 17,560 ha; equalling 18 m³ per hectare. The annual fellings were assumed to be as follows:

5,800 m³ of okoumé,
155,000 m³ of other main tree species,
132,000 m³ of secondary tree species,
11,200 m³ of other tree species.

From the economic point of view okoumé is the tree species of the greatest interest. Assuming that the average annual diameter increment of this tree species is 0.8 to 1.0 cm, 20 to 25 years are required for the transition of the average mean tree from one diameter class, e.g. 40–60 cm, into the next higher diameter class, e.g. 60–80 cm.

Not every tree in the stand grows to maturity. In establishing the logging cycle it was assumed that for the okoumé species in the 40–60 cm diameter class, one and a half trees were required for one tree to pass from this diameter class to the 60–80 cm diameter class over a period of 20 years. Or that 2 trees of okoumé of the 20–40 cm diameter class are required in order to obtain the same result; namely for one okoumé tree to grow from the 20–40 cm diameter class to maturity over a 40-year time span.

It was assumed that in the Sibiti-Zanaga inventory region a logging cycle of 20 years could be sufficient for felling mature timber for the national economy of the Republic of Congo, both for okoumé and for other commercial tree species.

The Sibiti-Zanaga inventory region is rich in timber resources, mainly okoumé. This growing stock makes it possible to expand timber exploitation even with increasing demand. Well organized forest regeneration after logging must ensure the lasting production of valuable tree species.

4.2.5 Forest inventory and management in the Kouilou region

In agreement with article 29 of the Forest Law No. 004/74 the Ministry of Forestry of the Republic of Congo has established eleven (Nos 1–11) management units in the southern part of Congo. This has been done in order to regulate and organize the exploitation of those forest tree species which may be economically utilized both for local processing and for export (Setzer, 1986).

For the rational utilization of the resources of natural forests in the southern part of Congo, at the beginning of 1981 an agreement was signed by the representatives of the Congolese Government, the United Nations Development Programme and the United Nations Food and Agriculture Organization concerning the realization of the project PRC 80/005 "Forestry Development". This related basically to an inventory of natural forests in the forest units Nos 2, 5 and 6 which are situated in the northwestern part of the forest sector of southern Congo (see map, Fig. 34).

The region in question refers to forests interspersed by bush and grassland savanna. Forested mountain slopes are intersected by valleys with watercourses and numerous lakes. In places, solid ground changes into swamps and periodically flooded land. About 80% of the investigated territory is covered by formations of the moist evergreen and semi-deciduous forest where commercial tree species such as *Aucoumea klaineana, Terminalia superba, Pycnanthus angolensis, Staudtia stipitata, Pterocarpus soyauxii* and *Entandrophragma angolense* occur frequently. Among the category of less important tree species, the most frequent are: *Ricinodendron africanum, Piptadeniastrum africanum, Celtis brieyi, Tessmania lescrauwaetii* and *Klainedoxa gabonensis*. In the lower shrub storey of forest stands

96

Annona arenaria, Bridelia ferruginea and *Hymenocardia acida* are found, and in the herbaceous groundcover *Cymbopogon giganteus, Eragrostis superba* and species of the *Andropogon* and *Hypparhenia* genera occur.

The climate is tropical, moist and hot with an average annual precipitation of about 1,500 mm. The rainy season lasts from October to June. At this time, the temperature exceeds 30°C. In the dry season the average temperature varies between 23 and 25°C. The relative air humidity does not sink below 75%.

The forest inventory over an area of about 1,100,000 ha was carried out during 1983 to 1987 by Congolese inventory teams under the supervision of FAO experts. The inventory was made by systematic sampling on line plots which were orientated in a north–south direction, consequently 9 or 189° to the magnetic pole. The inventoried plots on the inventory lines were of rectangular shape measuring 200 × 25 m = 0.5 ha. Sampling intensity (tree cruising) was done at 0.45% of the inventoried area on the assumption that the exploitable timber volume shall be determined with an accuracy of ±15% at a 5% level of significance.

After considering the possibilities of utilizing timber from the forests in the inventoried region both on domestic markets and abroad, the tree species were divided into two categories: the category of "main tree species" and the category of "remaining tree species". The list of main tree species is given below.

No.	Latin name	Commercial name	Family
1	*Aucoumea klaineana*	okoumé	Burseraceae
2	*Terminalia superba*	limba	Combretaceae
3	*Entandrophragma cylindricum*	sapelli	Meliaceae
4	*E. utile*	sipo	Meliaceae
5	*E. angolense*	tiama	Meliaceae
6	*E. candollei*	kosipo	Meliaceae
7	*Khaya anthotheca*	acajou	Meliaceae
8	*Lovoa trichilioides*	dibetou	Meliaceae
9	*Guarea cedrata*	bossé	Meliaceae
10	*Dumoria africana*	douka	Sapotaceae
11	*Baillonella toxisperma*	moabi	Sapotaceae
12	*Gambeya africana*	longhi	Sapotaceae
13	*Autranella congolensis*	mukulungu	Sapotaceae
14	*Oxystigma oxyphyllum*	tchitola	Caesalpinaceae
15	*Gossweilerodendron balsamiferum*	tola	Caesalpinaceae
16	*Dacryodes pubescens*	safoukala	Burseraceae
17	*Dacryodes buettneri*	ozigo	Burseraceae
18	*Pycnanthus angolensis*	ilomba	Myristicaceae
19	*Triplochiton scleroxylon*	ayous	Sterculiaceae
20	*Chlorophora excelsa*	iroko	Moraceae
21	*Staudtia stipitata*	niové	Myristicaceae

No.	Latin name	Commercial name	Family
22	*Pterocarpus soyauxii*	padouk	Papilionaceae
23	*Nauclea diderrichii*	bilinga	Rubiaceae
24	*Guibourtia arnoldiana*	mutenye	Caesalpinaceae
25	*Afzelia bipindensis*	doussié	Caesalpinaceae
26	*Pericopsis elata*	kokrodua	Caesalpinaceae
27	*Millettia laurentii*	wengé	Papilionaceae
28	*Swartzia fistuloides*	paurosa	Caesalpinaceae
29	*Combretodendron africanum*	essia	Lecythidaceae
30	*Rhodognaphalon brevicuspe*	alone	Bombacaceae
31	*Canarium schweinfurthii*	aiélé	Burseraceae
32	*Berlinia bracteosa*	ebiara	Caesalpinaceae
33	*Daniellia klainei*	faro	Caesalpinaceae

All other tree species (numbering about 150) recorded by the forest inventory were included in the category of the so-called "remaining tree species".

The inventoried trees were classified into 10 cm diameter intervals. The main tree species were measured beginning from a diameter breast height of 20 cm, the remaining tree species from a D.B.H. of 60 cm.

The minimum exploitable diameter breast height or diameter above the buttresses are determined in Congo for commercial tree species in agreement with the Decree to the Forest Law No. 84/910/1984. For the following tree species exploitation is permitted:
– bahia, ebonies and niové from the diameter of 40 cm up,
– movingui and olon from the diameter of 50 cm up,
– aiélé, bilinga, doussié, faro, limba, safoukala, oboto and tali from the diameter of 60 cm up,
– ayous, azobé, iroko and okoumé from the diameter of 70 cm up,
– acajou, agba, dibetou, douka, kosipo, kevazingo, moabi, padouk, sepelli, sipo, tchitola, tiama and zingana from the diameter of 80 cm up.

All other tree species can be exploited upwards from the minimum diameter breast height of 60 cm. The minimum exploitable diameter of tree species is determined according to the growth possibilities of the given tree species. The minimum exploitable diameters established by the state for the various tree species were taken into account in computing the allowable cut.

For the determination of the standing timber volume in the above-mentioned three management units, volume tariffs were used which were set up in previous forest inventories, or new volume tariffs were constructed for some main tree species.

The forest inventory in the management unit No. 6 – Divenie – was executed between 1983 and 1985. The forest management plan for this unit was prepared and approved in 1986. In the other two units field inventory work was terminated in 1987.

The management unit (L'unité forestière d'aménagement) No. 2, Madingo Kayes has an area of 510,580 ha. To the west it borders on the Atlantic Ocean, to the north it extends to the border with Gabon, to the south it abuts on the Kouilou river, and to the east it adjoins management unit No. 5, Kibangou. Administratively management unit No. 2 is included in the Kouilou region.

The landscape of the northern part of the management unit in inventory blocks a, b and c is mountainous with altitudes of 400–800 m. Its southern part (blocks a, e, f and g) is hilly with plateaus and wide valleys and altitudes of up to 150 m. Diverse soil types occur, from poor, shallow and gravelly soil in the Mayombe Mountains to fertile alluvia suitable for agriculture which is mainly found in the coastal region.

In the Kouilou region in the zone of management unit No. 2 there are approximately 16,550 human inhabitants and the average population density is 0.4 per km^2. In spite of this low population density a considerable part of the forest is degraded and systematically burnt for temporary agricultural crops (Setzer, 1986) (Fig. 47).

Fig. 47. Systematical burning down of the forest for temporary agricultural crops. Mayombe Mountains, Congo.

The total area of 510,580 ha of the management unit includes 398,980 ha of forests (78%), 95,856 ha of savanna (19%) and 15,744 ha of surface water and roads (3%). The natural closed tropical forest grows on 180,000 ha which equals 45% of the forest area, and the secondary closed tropical forest occurs on 157,074 ha representing 39% of the forest area. The remaining 16% of forest includes forest on degraded soil, on swamps or on periodically inundated land. Grassland and bush savanna are concentrated in the blocks e, f and g and which represents 25%, 40% and 27%, respectively, of the area of each block. Degraded forest occurs mainly in the neighbourhood of settlements and along roads.

Before starting the inventory work, management unit No. 2 was divided into seven inventory blocks with area between 31,250 and 104,750 ha. The inventory blocks will eventually have the function of logging blocks (Setzer, 1986).

Numerical data obtained from the forest inventory in southern Congo were processed according to the "FIDAPS" standard programme for the complex automated processing of inventory results from natural tropical forests prepared by the Forest Department of the FAO in Rome. Using "FIDAPS" formulae were also computed for setting up volume tariffs for economic (commercial) tree species in Congo.

Among the main growth indices of high tropical forests are average tree numbers converted to the area of 1 ha. Average tree numbers classified by diameter classes with an interval of 10 or 20 cm are usually given for the most important commercial tree species, or for categories of tree species. The average number of trees per hectare for the main tree species by diameter classes, and the average number of mature trees per hectare for the other tree species are given for the various inventory blocks and for the entire management unit No. 2 in Table 23.

Table 23. The mean number of trees per hectare. Forest management unit No. 2

Inventory blocks	Main species					Other species 60+ cm
	Diameter breast height classes in cm					
	20–40	40–60	60–80	80+	Total	
a	7.04	5.27	3.55	2,70	18.56	12.33
b	2.26	2.44	1.45	1.32	7.47	13.05
c	4.06	4.40	3.92	2.41	14.79	16.47
d	6.57	4.31	3.02	1.85	15.75	14.21
e	4.31	2.65	1.68	0.71	9.35	11.87
f	6.76	4.38	2.11	0.84	14.09	9.84
g	3.03	2.31	1.30	0.77	7.41	11.95
Management unit	4.40	3.57	2.44	1.54	11.95	13.48

Data presented in Table 23 indicate a great range in the average number of trees per hectare for the main and other tree species in the various blocks. The inventory blocks b and g are the poorest with 7 trees per hectare on the average for the main tree species, and 12–13 mature trees per hectare for the other tree species. The largest average number of trees for the main tree species occurs in block a (18 trees per hectare) and in block d (16 trees per hectare). The largest representation of the category of other tree species is in block c (16 mature trees per hectare). In the whole of management unit No. 2 there were on average 12 trees per hectare of the main tree species, including 4 mature trees. The average number of mature trees of the so-called "other tree species" was 13 per hectare, and the largest representation of species among this category include *Parinari glabra*, *Tessmannia lescrauwaetii*, *Uapaca guineensis*, *Monopetalanthus letestui* and *Pentaclethra eetveldeana*.

Niové, with an average number of 1.7 mature trees per hectare, was the most frequent commercial tree species in management unit No. 2. The smallest average number of exploitable niové trees occurred in inventory blocks a̱ and ḇ (1.2 trees per hectare), the largest average number of exploitable trees (2.8 per hectare) was found in block f̱. Okoumé occurred in all inventoried blocks, from 0.2 exploitable trees per hectare (block e̱) to 1.8 exploitable trees per hectare (block a̱). Limba occurred in blocks a̱, c̱ and d and varied from 0.1 to 0.6 exploitable trees per hectare on average. From among other main tree species safoukala, bilinga, longhi and ozigo were the most frequent.

Another important growth index in natural high tropical forests is the average growing stock, mainly of commercial tree species, given in m³ per hectare. In order to compute the average growing stock in·m³ per hectare and the growing stock in the various blocks or in the management unit, volume tariffs were used for each tree species adopting mathematical functions, e.g. for:

okoumé	$V = 8.18D^{1.81}$,
limba	$V = -0.0265 + 10.725D^2$,
niové	$V = 11.63D^{2.114}$,
safoukala	$V = 10.76D^{2.169}$,
iroko	$V = 0.4023 + 7.2904D^2$.

The average growing stock of mature trees for the main tree species in each inventory block and in the whole management unit is given in Table 24. The largest average growing stock of exploitable trees in the whole management unit occurred with the tree species niové: 6 m³ per hectare. For this tree species the lowest average exploitable growing stock occurred in blocks a̱ and ḇ (4 m³ per hectare) and the highest in block f̱ (11 m³ per hectare). The average growing stock of mature trees of okoumé in the management unit amounted to 4 m³ per hectare; being lowest in inventory block e̱ (1 m³ per hectare) and at its peak in block a̱ (13 m³ per hectare). The average growing stock of mature trees for all main tree species in the management unit was 28 m³ per hectare; being lowest in inventory block g̱ (circa 16 m³ per hectare) and highest in blocks a̱ and c̱ (ca 42 m³ per hectare).

From the commercial point of view, only the growing stock of exploitable mature trees of the main species can be considered. There is at present no demand on the domestic and foreign markets for the timber of other tree species. It may be stated that in management unit No. 2 (a production forest with an area of roughly 400,000 ha), the forest inventory established that there was over 11 million m³ of marketable timber. Okoumé, limba and niové – which are most in demand in southern Congo – contributed to 40% of the exploitable growing stock (Fig. 48). The largest exploitable timber stock of okoumé occurred in blocks a̱ (about 520,000 m³), c̱ (about 505,000 m³) and ḇ (about 360,000 m³). As for limba, the largest amount of exploit-

Table 24. The mean exploitable growing stock in m³ per hectare of main tree species. Management unit No. 2

Species	Inventory blocks							Manage-ment unit
	a	b	c	d	e	f	g	
niové	3.7	4.3	5.8	4.7	8.3	10.9	6.1	6.2
okoumé	12.8	4.4	5.5	1.9	1.2	2.5	1.8	4.3
safoukala	3.8	1.0	5.8	7.6	3.8	0.3	1.1	3.1
bilinga	3.5	2.4	4.7	1.5	–	0.5	0.1	2.1
ilomba	2.2	1.7	3.7	2.2	1.6	1.1	0.6	2.0
longhi	2.2	0.2	1.3	5.0	–	0.2	0.3	1.0
ozigo	4.0	2.2	0.5	0.1	0.1	–	0.1	1.0
limba	1.1	0.4	1.5	4.4	–	–	–	0.9
padouk	0.5	0.9	0.8	0.4	0.2	0.2	0.6	0.6
tola	1.2	0.1	0.8	–	0.1	–	0.2	0.4
kosipo	0.7	0.2	0.8	–	–	–	0.1	0.3
moabi	0.7	0.5	0.4	0.2	–	–	0.1	0.3
paurosa	0.2	0.2	0.2	0.1	0.3	0.6	0.3	0.3
dibetou	0.2	0.1	0.5	–	–	0.1	0.2	0.2
doussie	0.1	–	0.3	0.1	0.1	–	0.1	0.1
other	5.7	3.5	9.3	6.3	2.8	4.7	4.2	5.4
Total	42.6	22.1	41.9	34.5	18.5	21.1	15.9	28.2

able timber occurred in blocks c (about 138,000 m³) and d (131,000 m³). The largest amount of exploitable timber stock for niové occurred in blocks f (about 568,000 m³), c (about 533,000 m³), g (about 385,000 m³), b (about 352,000 m³) and e (about 327,000 m³). In inventory blocks f and g niové and okoumé, respectively, represented about 63% and 49% of the exploitable timber stock of all main tree species taken together.

From among commercial tree species occurring in the management unit No. 2, okoumé is of the greatest economic importance (Fig. 49). A considerable amount of logs and peeled veneer go for export. On the domestic market sawn wood production is more used. On world markets there is a constant demand for limba, both in the form of logs and veneer, cut or peeled. In the category of main tree species, niové is represented most. Its timber is heavy, hard and durable. It is used mainly for the production of sawn wood and parquetry. It is also much in demand for export.

The management of forest resources should be in full agreement with the economic and social development of the Republic of Congo, taking into account both the production function of the forest and its uncommercial benefits. Such a broad term of management of the Congolese forest massifs cannot be applied at present for objective reasons.

In management unit No. 2 (Madingo-Kayes) floristically rich natural moist high tropical forests occur. Here there are trees of various species of different ages, with

Fig. 48. Group of trees of *Terminalia superba* in a natural tropical forest. N'Gouha II, Congo.

Fig. 49. Increment experimental plot in a natural tropical forest. Marked trees of *Aucoumea klaineana*. N'Gouha II, Congo.

different space distributions, and with different dimensions; some trees attaining a diameter of 160 cm and a height of 50 m. In such natural conditions selective logging methods may be used for the regulation of fellings (Borota, 1987). According to this method, in a given time period mature and over mature trees, starting from the minimum exploitable diameter breast height for the commercial tree species, are marked for felling. An economic growing stock and a constant yield from the forest should be obtained by careful logging operations and ensuing species enrichment through afforestation by commercial tree species (Nair et al., 1985).

The logging cycle in the management unit was put at 25 years taking into account the established current annual diameter increment for the main commercial tree species, amounting, on average, to 0.8 cm. For example, an average tree of okoumé

Fig. 50. Shifting cultivation in a 35-year old plantation of *Terminalia superba*. Region of Mayombe, Congo.

with an annual diameter increment of 0.8 cm needs 25 years to pass from the 50 to 70 cm diameter class to the next higher exploitable diameter class.

In order to saveguard the natural regeneration of exploited tropical forests the assumption is usually made to retain in the diameter class of 20–60 cm at least 15 average trees of commercial species per hectare. Since in management unit No. 2 the average number of trees of commercial species per hectare varied from 5 to 12, it shall be necessary to carry out complementary afforestation in the management unit after logging operations. It is recommended to apply in this work well-tried experience obtained mainly in afforestation with limba (Fig. 50).

The allowable cut in the various blocks of the management unit was computed using the following simple equations:

Annual possibility by area = total forest area of the given block divided by the logging cycle. For example, for block a:

$$\frac{40,633 \text{ ha}}{25} = 1,625 \text{ ha},$$

which corresponds to an annual cut by timber volume of 69,236 m^3, this being the growing stock for the main tree species in block a divided by the cutting cycle (25 years).

Results of the forest inventory and management work showed that the exploitable growing stock of the main tree species in management unit No. 2 is sufficiently abundant (as is also the case in management units Nos 5 and 6) to meet the requirements of the national economy of the Republic of Congo both for domestic processing and export.

4.2.6 Production, processing and export of timber

Timber felling and processing is carried out in the Republic of Congo by organizations of the state sector, by mixed joint-venture companies, foreign firms and by private Congolese enterpreneurs. In 1986 foreign firms accounted for 56%, mixed joint-venture companies 20%, private Congolese enterpreneurs 15% and the state sector 9% of the total log production (715,000 m^3). According to present rules and regulations the logging companies must process in Congo at least 40% of felled timber. The remaining extracted logs are purchased from the logging companies and are either sold to the woodworking industry in Congo or are exported in the form of logs by a special state organization – the Congolese Timber Office (Office Congolais de Bois – O.C.B.) which has a monopoly for the purchase and selling of timber produced in Congo (Act No. 191/1975).

Information on the development of log production of commercial tree species in boom years (1969) and at present (1982, 1984, 1986) is given in Table 25. In the recent past mainly the commercial tree species of limba and okoumé were exploited

Table 25. Logs production of commercial timber species in Congo

Species	1969	1982	1984	1986
	m³			
Okoumé	391,200	184,100	194,300	199,000
Limba	161,700	11,300	14,900	36,300
Sapelli	42,200	115,200	174,400	277,200
Sipo	30,800	21,100	25,800	32,300
Agba	24,400	15,800	15,000	5,900
Moabi	21,100	11,800	10,200	18,100
Douka	20,100	7,200	9,500	9,200
Niové	19,700	15,500	26,300	33,400
Tiama	12,800	9,300	10,900	15,300
Tchitola	12,500	1,800	500	500
Khaya	10,700	800	4,200	4,400
Kokrodua	10,600	100	200	100
Wawa	8,300	6,600	3,500	16,100
Kosipo	7,500	4,100	3,000	4,800
Dibetou	6,700	1,100	3,500	4,500
Iroko	4,400	6,500	18,500	18,000
Longhi	3,700	8,300	11,500	9,500
Padouk	2,800	4,100	5,900	5,800
Bilinga	1,800	13,000	14,300	25,600
Doussié	1,000	1,100	1,300	2,100
Other	26,000	77,700	38,400	46,800
Total	820,000	516,500	586,700	714,900

in Congo. In 1969 these two tree species, made up 67% of the total log production of Congo. In 1986 their share in the total log production declined to 33%. Until 1965 limba was the most exploited tree species in Congo. Then okoumé ranked first until 1984 and from 1985 on sapelli took over the first place in timber production, being felled mainly in the northern part of Congo. Compared to 1969, the fellings of limba, agba, douka, tchitola, acajou (khaya) and kokrodua declined substantially whereas the exploitation of sapelli, niové, ayous (wawa), iroko, longhi and mainly bilinga increased. In 1985, the following tree species fetch the highest prices in Congo: kokrodua, wengé, sipo, sapelli and doussié (the free on board – f.o.b. – price of logs in Pointe Noire equals 1,500 to 1,900 FF per m³; FF = French Franc). Medium to high prices are realized by the following tree species: moabi, douka, longhi blanc, limba, okoumé, iroko, khaya, padouk, mukulungu, agba, dibetou and tiama (the f.o.b. price of logs equals 800 to 1,100 FF per m³). The cheapest tree species in Congo are movingui, izombé, bilinga, tali, abura, limbali, aiélé, olon, niové, kotibé, safoukala and ebiara (the f.o.b. price of logs equals 500–700 FF per m³).

Timber processing began on the territory of today's Congo at the beginning of the 20th century when small sawmills were established. At present sawmilling is also

106

the basic branch of the woodworking industry. In 1986 76,700 m^3 of sawn wood was processed in Congo. Foreign private firms took up 82% of the processing of sawnwood, domestic private enterpreneurs took up 9%, mixed joint-venture companies took up 7% and the state sector had 2%. Of the total annual amount of sawn wood 60–70% is consumed within Congo.

Veneer sheets and plywood are important forest products in Congo. Before independence was proclaimed in 1960 there was only one veneer sheet factory in the country with an annual production of about 8,000 m^3 of peeled veneer sheets. At present there are four factories in the country producing annually 50,000–60,000 m^3 of peeled veneer sheets and about 2,000–2,500 m^3 of sliced veneer sheets. The highest production of veneer sheets was recorded in Congo in 1973 (96,000 m^3). In 1975 veneer sheet production declined to 43,000 m^3 as a consequence of the world slump in the woodworking industry. The only plywood factory in the country manufactures 6,000–7,000 m^3 of plywood annually, mostly for domestic consumption.

There is a furniture-making joint-venture company in the country. Furniture is manufactured in small businesses for domestic requirements only. The construction of a pulp and paper mill in the region of Pointe Noire has not started yet although to this purpose a society for the cultivation of industrial plantations (Unité d'Afforestation Industrielle du Congo) using suitable eucalypt hybrids and tropical pine species has been established. The society annually produces 2 million plants. Up to now a green belt with an area of 25,000 ha has been established around the port of Pointe Noire.

Data on the development of log production and main forest products in Congo since 1962 are given in Table 26. The quantity of annual log production increased up to the beginning of the seventies. As a result of the steep price increases for tropical timber on world markets, as well as a consequence of drilling for oil in the Re-

Table 26. Trends in production of the main forest products in Congo

Year	Logs	Sawn wood	Veneer sheets	Plywood
	1,000 m^3			
1962	385	33	9	–
1964	635	29	22	–
1966	691	32	38	–
1968	777	42	60	–
1970	810	43	67	–
1972	771	48	91	–
1974	470	48	71	–
1976	400	60	55	–
1978	461	43	78	2
1980	603	64	75	4
1982	516	66	63	4
1984	587	60	65	6
1986	715	77	52	7

public of Congo, log production declined in 1976 to 49% of the amount production in 1970. Although since 1976 the exploitation of industrial wood in Congo has been on the increase, the maximum log production obtained in 1986 represented only 88% of the production in 1970. The amount of development for the production of sawn wood, veneer sheets and plywood does not reflect the possibilities which exist in this country's rich timber resources.

Timber export plays an important role in the national economy of the Republic of Congo. In the early seventies timber exports represented up to two thirds of the value of total exports. However, since 1975 oil extracted mainly from coastal regions has played the main role in the Congolese national economy.

Logs have been up to the present time the most important form of timber exported. The development of the export of logs, sawn wood and veneer sheets from Congo is given in Table 27. As shown in the table, the export of tropical logs in the period after the war displays a tendency for increase until the early seventies when the export reached a volume of 600,000 m^3. Afterwards log export declined dramatically. In 1976 only 144,000 m^3 of logs were exported, i.e. 24% of the export volume of 1970. In the beginning, the amount of log export was very high in comparison with total log production, reaching 93% in 1962. Later, log export declined until between 1976 and 1978, when 36% of the total was exported. At present about 40% of the total log production is exported (1986). Compared to the years of the highest boom in timber commerce, 40–50% of the then total volume is now exported.

Table 27. Export of the main forest products from Congo

Year	Logs	Sawn wood	Veneer sheets
	1,000 m^3		
1962	360	11	9
1964	537	6	17
1966	556	5	36
1968	570	8	55
1970	592	12	67
1972	473	16	81
1974	277	24	48
1976	144	16	55
1978	167	18	77
1980	281	37	67
1982	206	24	59
1984	250	28	59
1986	287	23	47

Until 1965 limba was the most exported tree species making up 70% of the logs exported (1960). After 1966 okoumé became the most exported tree species, its share in the total log export represented 52% in 1970. In 1985 sapelli, okoumé, tiama, sipo, iroko and limba are the species most frequently exported from Congo in the form of logs. About 70% of all logs exported belongs to these six species. From among lesser-known species longhi, bilinga, kanda, mukulungu, movingui, abura, tali and faro are at present beginning to be exported in larger quantities (exceeding 1,000 m³ annually) in the form of logs.

Sawn wood export does not reflect the existing possibilities. Sawn wood is exported from Congo in the range of 20,000–30,000 m³ annually. In these exports sawn wood of niové, sapelli, ayous, bilinga and sipo predominates.

The Republic of Congo together with the Ivory Coast are among the largest exporters of veneer sheets in Africa. The largest quantity of veneer sheets was exported from Congo in 1972 (91,000 m³). At present 45,000–60,000 m³ of peeled veneer sheets obtained mainly from okoumé, is exported from Congo each year.

France, Portugal, Italy, F.R.G., the U.S.S.R. and Spain are the largest importers of logs from Congo. About 80% of the logs exported from Congo goes to these six countries. France imports most of the okoumé logs. The largest quantity of sawn wood is imported by Rèunion, mainly niové (11,700 m³ in 1986), followed by Spain, F. R. G. and France. France, F.R.G., Tunisia and the Netherlands are the largest importers of peeled veneer sheets from Congo, and account for about 70% of the export (1986).

The Government of the Republic of Congo makes every effort to develop forestry and to establish an adequate woodworking industry. The Government is interested in economic cooperation with foreign business partners who have a good command of the technology of wood processing and may assist in selling forest products on world markets.

4.3 Republic of Gabon

4.3.1 Basic data

The Republic of Gabon (République Gabonaise) is situated in the region of the Gulf of Guinea along both sides of the equator between 2°20′ N and 4° S latitude and between 8°40′ and 14°30′ E longitude. Its western natural border, having a length of 800 km, is formed by the Atlantic Ocean. To the north Gabon borders on Equatorial Guinea and Cameroon and to the east and south Gabon borders on the Republic of Congo.

The country has a surface area of 267,000 km² and is administratively divided into 9 provinces. The population is about 1.1 million (1984). The largest part of the population belongs to the northwestern group of the Bantu people. Since 1964 Gabon has been a member of the Customs and Economic Union of Central Africa – UDEAC (L'Unité Douanière et Économique de l'Afrique Centrale). Chad, Cameroon, Congo and the Central African Republic are also included in the UDEAC.

The capital and largest port is the city of Libreville (about 200,000 inhabitants). The second most important city of the country is harbour Port Gentil (100,000 inhabitants). The city of Lambaréné on the banks of the Ogooué river is well known for its Hospital and Research Institute for Tropical Diseases (founded by Dr. A. Schweitzer). The official language is French, the currency unit is the CFA Franc.

The coastal lowland belt of Gabon is formed by swamps and mangrove forests. In the north of the country the coastline is dissected. To the east and northeast the landscape turns into plateaus and mountains with the highest altitude being the Monts de Crystal (about 1,000 m above sea level). The mountain massif Chaillu with its highest peak of Nimi (1,370 m above sea level) is situated to the south of the Ogooué river. The Ogooué river with its affluents the Ivindo and the N'Gounié, and the Nyanga river are the most important watercourses. The rivers are navigable all the year round only on their lower courses.

Gabon belongs to the woodiest African countries having a forest percentage of 80%. In the basin of the Nyanga river in the southern part of the country there are extensive secondary savannas covering 15% of the country's territory. In the lake region there are swamp forests and on the littoral and in river deltas mangrove forests occur.

The climate is tropical, equatorial and moist. The average temperature of the warmest month – April – is 25–28° C, of the coolest month – July – 20–23° C. The temperature difference between night and day does not exceed 10° C – even on mountain sites. Relative air humidity varies from 75 to 90%.

Annual precipitation is distributed unequally. In Cocoleach in the northwestern part of the country the annual rainfall is 3,500 mm, in Libreville it is 2,680 mm and in Port Gentil it is 1,950 mm. In the south and towards the interior, precipitation diminishes. In Mayombe the average annual precipitation amounts to 1,770 mm and in Tchibango it equals 1,400 mm. In Gabon two rainy seasons and two dry seasons alternate. The first rainy season lasts from the middle of January until the middle of May and the second lasts from early October to the middle of December.

Gabon is an agricultural country with a developing industry. About 70% of the population subsist on agriculture. For food crops mainly manioc (cassava), sweet potatoes, rice, maize and bananas are grown. Industrial crops include coffee, cocoa, rubber, peanuts and palm oil. The fisheries of the country are important. The raising of cattle is made impossible because of tsetse fly. Oil drilling in the littoral near Port Gentil, and the extraction of manganese and uranium ore are the most important branches of the developing industries. The woodworking industry, an oil refinery and food processing are the basic pillars of the processing industry. About 15% of the population is employed in mineral extraction and food processing.

110

4.3.2 The types of forest formations and the categories of forest tree species

About 80% of Gabon's landscape is covered by evergreen or semi-deciduous moist tropical forests with a very rich vegetation and fauna. De Saint-Aubin (1963) classified the Gabonese tropical forests into the following ecosystems:

– Coastal forest region with the *Aucoumea klaineana – Dacryodes buettneri* and the *Aucoumea–Sacoglottis* communities,
– Southern forest region of the *Aucoumea–Dacryodes* community with an admixture of *Terminalia superba*,
– Central and eastern forest region extending to the south of the Ogooué river; it includes the *Aucoumea–Dacryodes* forest community with an admixture of *Scyphocephalium ochocoa* and *Paraberlinia bifoliolata* tree species,
– Central and eastern forest region extending to the north of the Ogooué river. In the Crystal Mountains the *Aucoumea–Dacryodes–Monopetalanthus* community occurs. In the region between the Crystal and the M'Voung Mountains *Gossweilerodendron balsamiferum*, *Scyphocephalium ochocoa* and *Paraberlinia bifoliolata* are more represented. In the forests to the east of M'Voung *Aucoumea klaineana* does not occur but *Scyphocephalium ochocoa* and *Paraberlinia bifoliolata* are represented.
– Northern forest region where *Aucoumea klaineana* and *Scyphocephalium ochocoa* do not occur. Of greater occurrence are the species *Terminalia superba*, *Triplochiton scleroxylon* and species of the *Entandrophragma* genus.

In the Gabonese forests okoumé *(Aucoumea klaineana)* is the most important commercial tree species. With regard to the occurrence of this tree species the forests of Gabon are classified from a practical point of view as follows:

(A) Forests on solid soil

1.	(a) with okoumé, logged	7,880,000 ha,
	(b) with okoumé, not yet logged	6,515,000 ha,
2.	(a) without okoumé, logged	20,000 ha,
	(b) without okoumé, not yet logged	5,970,000 ha
	Total	20,385,000 ha

(B) Swamp and mangrove forests   1,095,000 ha

    Grand total   21,480,000 ha.

The grassland and bush savanna extends over an area of more than 3 million ha.

A special type of the tropical forest is the mangrove. It occurs on the coast near river deltas. The dominant tree species is *Rhizophora racemosa*. The trees seldom exceed a height of 15 m.

In addition to natural tropical forests there are in Gabon 25,000 ha of forest plantations, mainly under okoumé.

According to the Forest Act the country's forests are classified, with regard to logging and timber utilization, into three forest zones (Fig. 51).

The first zone has an area of about 5 million hectares. It includes a 50–250 km wide coastal belt with the exception of the southeastern part of the country. In the first forest zone the ecosystem of the *Aucoumea–Dacryodes* community dominates with *Aucoumea klaineana* prevailing. Among other tree species *Dacryodes buettneri*, *Sacoglottis gabonensis*, *Irvingia oblonga*, *Calpocalyx heitzii*, *Klainedoxa gabonensis*, *Plagiostyles africana*, *Paraberlinia bifoliolata*, *Guibourtia ehie* and species of the *Dialium* and *Monopetalanthus* genera and others occur frequently.

GABON

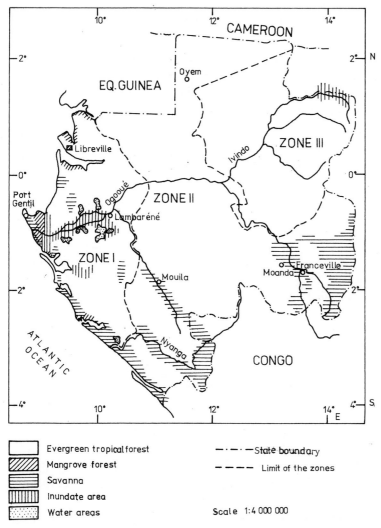

Fig. 51. Orientation map of the Republic of Gabon.

Forests of the first forest zone have been logged since the beginning of the 20th century, especially in the vicinity of watercourses. Okoumé was the main object of logging. Management is based on the priority to be given to okoumé, a species of great commercial interest. Logging and regeneration operations are to be planned in harmony so as to ensure sustained yield (Catinot, 1978).

The second zone is the largest one, with an area of around 15.3 million ha. It is situated between the first and third zones. It is varied as to natural conditions as it includes seven natural regions. In the second zone, okoumé is absent in certain parts of the forest massif. The Government supports the transfer of logging operations into the second zone by granting financial incentives to logging and wood processing companies.

The third zone includes forests in the northeastern part of the country with an area of about 7 million ha. Forests both with and without okoumé are included. So far in this zone industrial logging has not taken place because of lack of access.

Tree species growing in Gabon are classified into three groups A, B and C according to their importance and utilization. The group A includes 18 tree species with the greatest economic importance for industrial processing and export. These are the so-called "commercial tree species".

The group B includes the so-called "secondary", or, still better, the "lesser-known tree species". These are 17 tree species which are used locally for industrial processing; they are exported to a smaller degree. It is assumed that these tree species, because of their favourable technological properties, will be used more in the near future.

All other tree species growing in the Gabonese forests are included in the group C; they are little used for local industrial processing or not used at all because the properties of their timbers are not known sufficiently. The group C includes more than 250 different species (Développement forestier, Gabon, rapport, FAO, 1970).

The list of tree species is given below.

Group A – Commercial tree species:

Aucoumea klaineana PIERRE – okoumé

Khaya ivorensis A. CHEV. – acajou

K. anthotheca C. DC. – acajou

Gossweilerodendron balsamiferum HARMS – agba, tola

Guarea cedrata PELLEGR. – bossé

Lovoa trichilioides HARMS – dibetou

Dumoria africana A. CHEV. – douka

Afzelia bipindensis HARMS – doussié

Pycnanthus angolensis WARB. – ilomba

Chlorophora excelsa BENTH. et HOOK. f. – iroko

Entandrophragma candollei HARMS – kosipo

Baillonella toxisperma PIERRE – moabi

Tarrietia densiflora AUBR. et NORMAND – niangon

Fagara heitzii AUBR. et PELLEGR. – olon
Dacryodes buettneri (ENGL.) LAM. – ozigo
Entandrophragma cylindricum SPRAGUE – sapelli
E. utile SPRAGUE – sipo
E. angolense C. DC. – tiama
Oxystigma oxyphyllum J. LÉONARD – tchitola
Group B – Lesser-known tree species:
Poga oleosa PIERRE – afo, ovoga
Canarium schweinfurthii ENGL. – aiele
Paraberlinia bifoliolata PELLEGR. – awougha, beli
Lophira alata BANK ex GAERTN. f. – azobé
Mitragyna ciliata AUBR. et PELLEGR. – bahia
Nauclea trillesii MERR. – bilinga
Guibourtia demeusei J. LÉONARD – ebana, bubinga
Coelocaryon klainei PIERRE – ekoune
Testulea gabonensis PELLEGR. – izombe
Guibourtia tessmannii J. LÉONARD – kevazingo
Nesogordonia papaverifera R. CAPURON – kotibé
Distemonanthus benthamianus BAILL. – movingui
Staudtia stipitata WARB. – niové
Swartzia fistuloides HARMS – oken, pau rosa
Dacryodes igaganga AUBR. et PELLEGR. – ossabel
Guibourtia ehie J. LÉONARD – ovangkol
Pterocarpus soyauxii TAUB. – padouk
Group C – Other tree species:
Bombax chevalieri PELLEGR. – alone
Monopetalanthus heitzii PELLEGR. – andoung
Copaifera religiosa J. LÉONARD – anzem
Brachystegia laurentii LOUIS – bomanga
Piptadeniastrum africanum BREMAN – dabema
Berlinia spp. BENTH. et HOOK. f. – ebiara
Cylicodiscus gabonensis HARMS – edoun, okan
Didelotia letouzeyi PELLEGR. – ekop, gombé
Alstonia congensis ENGL. – emien
Ricinodendron africanum MUELL. ARG. – essessang
Klainedoxa gabonensis PIERRE ex ENGL. – eveuss
Vitex pachyphylla BAK. – evino
Ceiba pentandra GAERTN. – fromager
Daniellia klainei PIERRE ex A. CHEV. – lonlaviol, faro
Chrysophyllum spp. LINN. – longhi
Irvingia grandifolia ENGL. – olene
Newtonia leucocarpa GILBERT et BOUTIQUE – ossiniale
Symphonia globulifera LINN. f. – ossol, manil

Scyphocephalium ochocoa WARB. – sogho, soro

Erythrophleum ivorense A. CHEV. – tali

In the group C list only those species which could acquire economic importance in the near future are mentioned. According to the opinion of Gabonese foresters more than 70 autochthonous tree species may be used for industrial processing.

In the category of light weight wood species okoumé and ozigo are used mostly for peeled veneer sheets and cabinet making; ozigo is heavier than okoumé. The sequence in utilization, with regard to the suitability of timber and its technological properties, continues with aiélé, ilomba, limba, olon, tchitola, tola, emien and essessang. Peeled veneer sheets can also be obtained from tree species of the group C such as: ekop, bomanga, ebiara, faro, anzem, andoung, fromager and alone. Afo, bahia, sogho and ekoune are suitable both for cabinet making and peeled veneer sheets.

In the group of medium-hard wooded tree species acajou, sipo, kosipo, tiama, bossé, sapelli, dibetou and niangon are of general use. Douka, moabi, doussié, ossol, bilinga, izombé and movingui are suitable for producing decorative veneer sheets.

Hard and heavy weighted wood species, suitable for both decorative veneer sheets and construction, are valued as to quality according to the following sequence: bubinga, kevazingo, ovangkol, kotibé and oken. For railway sleepers, parquetry and also for luxury furniture the use of azobé, padouk, niové, tali, dabema, ossiniale, eveuss, miama and oken is recommended.

Okoumé is at present in Gabon the most widely represented, utilized and exported commercial tree species (Fig. 52). Its timber is durable and when dry, insect resistant, however, it is not very resistant against dynamic and static strain. Okoumé occurs most frequently at altitude between 200 and 500 m on medium rich soils where it attains a height of 30–50 m and a diameter breast height of 150 cm. The heartwood is of a rosy colour with a silky lustre.

Among other tree species frequently used in Gabon that of ozigo – *Dacryodes buettneri,* locally called assia, should be mentioned. It occurs frequently together with okoumé, but its natural distribution area is larger.

Ozigo grows to heights of 20–30 m and to a diameter breast height of 150 cm. It forms straight, cylindrical stems. The timber is of a greyish-rosy colour, resembling okoumé, but heavier, more elastic and stronger. The heartwood is more durable but the sapwood after felling is quickly infested by insects and fungi. Ozigo is the second most exported tree species in Gabon. It is used for peeled and decorative veneer sheets, plywood, parquetry, for cabinet making and in construction.

The third most used and exported tree species in Gabon with reference to quantity is ilomba – *Pycnanthus angolensis,* locally called eteng. The tree attains a height of 35 m and a D.B.H. of 120 cm. The timber is of a rosy-brown colour with coarse, parallel fibres and it is less durable than that of the previously mentioned species. It is used for plywood and blockboard, for surface veneer sheets, packing, for cheap furniture and in construction.

From among the category of lesser-known and other tree species andoung, lon-

115

Fig. 52. Landing of okoumé logs in the forest at the roadside. Region of Franceville, Gabon.

laviol and gheombi may be used for manufacturing peeled veneer sheets; awougha, obana and kevazingo for the production of decorative veneer sheets; ekoune and sogho for joinery; and movingui, padouk and bilinga for parquetry and joinery. Very hard timbers, such as azobé, alep, edoun, eveuss and olene can be used for the production of railway sleepers, for flooring and piles.

4.3.3 Results of forest inventories in the northern regions of the First Forest Zone

At present the Gabonese Government pays most attention to the little explored and inaccessible forests of the Second and Third Forest Zones with a view to the possibility of the rational utilization of their timber resources. However, from the economic point of view the First Forest Zone in the coastal region will always remain important for Gabon.

During his stay as a forestry expert in West Africa the author availed himself of the opportunity to study certain data from forest inventories which were carried out in 1962–1963 by the company "La Cellulose du Gabon" in the northwestern parts of the country. The inventoried forests were situated on clayey soil to the southeast of Libreville up to the Ogooué river (= Gabon I), to the north of Libreville (= Gabon II) in the forest regions of the Monts de Crystal, and on sandy soil around N'Dombo which lies to the east of Libreville (= Gabon III). Alle these inventoried regions are included in the First Forest Zone.

More than 100,000 ha of forest were inventoried using the method of sample inventory plots which were 1,500 m long and 30–40 m wide. The inventory covered 1 to 1.33% of the forest's area. Tree species of economic interest were callipered from the lower limit of the 40 cm diameter class and were recorded in diameter classes of 10 cm intervals (35 to 45 cm, 45 to 55 cm, etc.). Before starting the inventory work, aerial photographs were used for determining the forest boundaries and surface area.

Empirical values of the diameter distribution of trees of the most represented economic tree species and their calculated exponents, as well as the multiplication constant of the beta function, are given in Tables 28 and 29 for the location of Gabon I and Gabon II. Tree numbers are converted to an area of 1,000 ha.

Table 28. Actual data of diameter distribution of trees of the most abundant economic species and their calculated exponents and constant by beta function. Gabon I, area: 1,000 ha

D.B.H. classes	Timber species						
cm	Okoumé	Ozigo	Ilomba	Olon	Ekoune	Niové	Ovangkol
40	874	388	150	42	605	535	218
50	860	274	140	40	382	314	180
60	718	187	138	25	186	152	134
70	508	173	110	16	87	64	94
80	256	102	55	11	34	22	27
90	121	56	23	2	11	4	5
100	66	56	16	1	3	1	1
110	43	20	4	–	–	2	–
120	28	18	2	–	–	1	–
Total	3,474	1,274	638	137	1,308	1,095	659
α	0.2709	−0.1206	0.3027	0.2015	−0.0392	0.0269	0.2737
β	4.0489	1.7640	2.5954	2.6813	2.8763	5.0020	2.7103
Const.	0.0000	0.0149	0.0001	0.0000	0.0004	0.0000	0.0002

Volume tariffs were set up for the calculation of the growing stock of economic tree species. For okoumé the volume tariff was constructed according to the equation:

$$V = 1.045 + 0.0913\ D^2,$$

117

Table 29. Actual data of diameter distribution of trees of the most abundant economic species and their calculated exponents and constant by beta function. Gabon II, area: 1,000 ha

D.B.H. classes	Timber species						
cm	Okoumé	Ozigo	Ilomba	Olon	Ekoune	Niové	Ovangkol
40	726	576	187	60	530	1386	214
50	681	371	156	35	341	422	194
60	560	273	96	14	216	148	118
70	440	223	91	12	115	24	72
80	354	118	46	9	36	4	41
90	209	81	10	5	9	2	24
100	62	76	4	–	2	–	12
110	–	21	2	–	–	–	4
120	–	16	2	–	–	–	2
Total	3,032	1,755	594	135	1,249	1,986	681
α	0.0519	−0.1340	0.1566	−0.2327	0.0450	−0.0511	0.1292
β	0.9594	1.9193	3.0433	1.7382	2.8012	4.8897	4.3357
Const.	1.2599	0.0110	0.0000	0.0050	0.0004	0.0000	0.0000

where V is the volume of the stem in m^3, D the stem diameter at the height of 1.30 m, in cm.

After the termination of field inventory work in the forests of the First Forest Zone, the growing stock was established for the following tree species:

okoumé	13.8 million m^3,
ozigo	6.8 million m^3,
bahia	6.3 million m^3,
douka and moabi	2.0 million m^3,
ilomba	1.9 million m^3,
sogho and ekoune	1.9 million m^3,
ovangkol, ebana and kevazingo	1.7 million m^3,
bilinga, movingui and izombé	1.2 million m^3.

The average volume of exploitable timber for commercial tree species in the First Forest Zone was 10 m^3 per hectare. The average timber stock of all tree species from a diameter breast height greater than 10 cm is estimated at 200–220 m^3 per hectare in forests logged in the past, and about 250 m^3 per hectare in forests not logged as yet.

Also in the Second Forest Zone forest inventories have been carried out. In this zone the growing stock is estimated at about 40 million m^3 for okoumé, 14 million m^3 for ozigo, 16 million m^3 for ilomba, and about 5 million m^3 for douka and moabi.

The forests in the Second Forest Zone are heterogeneous. In the north of this zone, on an area of about 1.3 million ha, there are semi-deciduous, partially degraded forests. Okoumé, ozigo and sogho do not occur there. Limba and ilomba are represented rather frequently.

Reconnaissance inventory work has also been done in the Third Forest Zone in the northeastern part of the country. This is an extensive region of moist evergreen and semi-deciduous tropical forest where okoumé is absent. Taking into account all tree species with a diameter of 70 cm and more, the growing stock was computed in m^3 per hectare for trees mature enough for felling. From among 18 commercial tree species the largest exploitable growing stock per hectare included wawa (ayous) with 14.5 m^3, limba with 3.0 m^3, ilomba with 2.8 m^3 and ozigo with 2.3 m^3. From among other tree species sogho with 7 m^3 and andoung with 5 m^3 per hectare were the most productive.

On all three localities of the First Forest Zone okoumé has the largest representation and ozigo is a less frequent species. As for export, ilomba and olon are well represented. The so-called African mahoganies: sipo, kosipo, tiama, sapelli and khaya, as well as other commercial species are of rare occurrence. From among lesser-known species ekoune, niové, ovangkol, bahia and padouk in the locality of Gabon I are the most frequent, as are niové, ekoune, azobé, ovangkol and ossabel in the locality of Gabon II, while niové, awougha, ekoune, ovangkol, azobé and padouk in the locality of Gabon III are the most frequent species.

From among the category of other tree species the locality Gabon I included mostly alep, sogho, miama, eveuss, ebiara and andoung; the locality Gabon II included alep, gheombi, sogho, eveuss, miama and ekop; and the locality Gabon III included sogho, dabema, alep, gheombi, faro, bomanga and andoung.

Inventory data obtained from the investigated localities indicate a declining tendency of diameter distribution of trees with increasing tree diameter. In order to identify the curves of diameter frequency of trees in all three localities, i.e. Gabon I, II and III, the beta function was used by the author. This method was used with success in the evaluation of similar inventory data from high tropical forests in Ghana and Congo. Compensated curves according to the beta function for okoumé show for all three localities approximately the same course – an inverted, elongated "S" (Fig. 53). For other tree species, computed distribution curves of tree frequencies reveal a declining tendency from the smallest to the largest tree diameters.

Okoumé is a tree which grows to large dimensions, attaining a diameter of 150 cm and more. In the inventoried regions of Gabon, okoumé trees grew to a maximum diameter of 130 cm. The curves of diameter distribution of okoumé trees in all three localities are also distinguished by a relatively smaller number of trees in the two smallest diameter classes. This observation refers to localities where in the past intensive logging, mainly of okoumé, occurred. Although only trees of large diameter were felled, the logging operations apparently also destroyed smaller trees and the natural regeneration. This may explain the unusual course of the curves of diameter distribution of the okoumé trees.

There are about 20,000 ha of plantations with *Aucoumea klaineana* in Gabon. The plantations were established after 1945 usually in a spacing of 4×4 m using

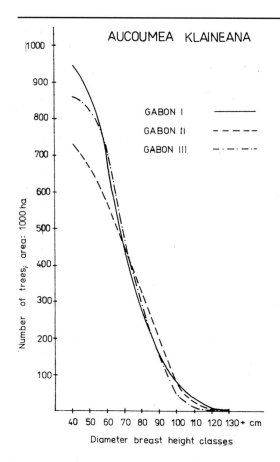

Fig. 53. Diameter distribution of trees for okoumé according to the beta function. Inventory regions Gabon I–III.

AUCOUMEA KLAINEANA

GABON I ——————
GABON II – – – – –
GABON III –·–·–··–··

Number of trees, area: 1000 ha

Diameter breast height classes

vigorous plants. It was assumed that the trees would reach 70–75 cm D.B.H. in 50 years, at which time 100–120 trees per hectare should remain in the stand. Among other tree species mainly *Gmelina arborea, Terminalia superba* and *Nauclea diderrichii* are used in the establishment of plantations. Recently plantations of *Pinus caribaea, P. oocarpa* and eucalypt hybrids (Boden, 1964) are being established for pulpwood production.

4.3.4 Wood processing industries and timber export

Forests rich in timber resources play an important role in the national economy of the Republic of Gabon. With regard to the low human population and the development of a wood processing industry, Gabon is one of the few developing countries having a relatively high proportion of industrial wood (over 50%) in its annual timber production.

A practical illustration of the development of timber production in Gabon during the 20 years preceding 1985 is given by the data in Table 30. The data show that

Table 30. Production of roundwood in Gabon (Yearbook of Forest Products, FAO, Rome, 1975, 1986)

Year	Industrial roundwood		Fuel wood	Total
	1,000 m^3	%	1,000 m^3	1,000 m^3
1966	1,498	59	1,027	2,525
1968	1,639	60	1,071	2,710
1970	1,880	63	1,100	2,980
1972	2,270	67	1,124	3,394
1974	2,087	65	1,146	3,233
1976	1,174	50	1,166	2,340
1978	1,295	52	1,184	2,479
1980	1,347	53	1,207	2,554
1982	1,135	48	1,222	2,357
1984	1,484	55	1,222	2,706

there was an increasing trend mainly in the production of industrial wood until the early nineteen-seventies, i.e. until the dramatic increase of prices for tropical timber, since when the production of industrial wood declined in all developing countries. A further decline in the production of industrial wood occurred in 1982 at the time of the world economic recession. Consumption of fuelwood in the observed twenty-year period shows a slowly increasing tendency. The share of industrial wood in the total timber production increased during 1966–1972 from 59–67%. Parallel to the diminishing export of tropical logs the share of production of industrial wood also declined to 50% in 1976 and to 48% in 1982.

Gabon is among the leading processors and exporters of tropical timber in Africa. Organized logging began at the beginning of the 20th century. At present, the logging and processing of timber is carried out by both national and foreign companies. Today, as in the past, transportation methods including river navigation are the limiting factor or bottleneck for the expansion of logging.

In order to gain access to extensive forest regions of the Second and Third Forest Zones, the Government has built the Transgabon Railway from the port of Libreville to Franceville in the southeast, and to Belinga in the northeast of the country. In these so far inaccessible regions great possibilities for enterprise have opened up mainly for large logging and processing firms with foreign capital and adequate mechanization. The Government takes care of the construction of the main roads, but forest roads for timber hauling are constructed by the respective logging companies.

Since 1975 about fifteen large logging and industrial companies have acquired logging licences for 30 years on an area of forest exceeding 100,000 ha in these new regions. Their task, in addition to systematic logging and partial timber processing in the country, also includes reafforestation work for safeguarding the regeneration and production capacity of the exploited forest. The companies which own at least one wood processing plant in Gabon enjoy tax concessions. The state owned company "Société nationale des bois du Gabon" with monopoly powers is responsible for the purchase of timber for export and for the supervision of the export of tropical timber.

Until the nineteen-fifties, the main branch of wood processing in Gabon was saw-milling, mainly to meet local demand. At present, plywood production and the manufacture of furniture are mainly promoted.

In the nineteen-seventies, the wood processing industry entered a period of boom which resulted in the modernization of existing plants and in the establishment of new ones.

In 1976 the Government of Gabon decided to use part of the financial proceeds from oil exports as a source of investment for the development of the metallurgical, woodworking and food processing industries.

The promotion of the wood processing industry aims at its modernization, and to increase the share of lesser-known tree species both in processing and in the export of timber. Another aim is to cut down transportation costs, this is to be achieved by deploying wood processing plants all over the country, and also by constructing roads for timber transportation from the forest to processing plants and ports. The Government favours the establishment of mixed enterprises, and provides for customs, tax and other relief for both domestic and foreign enterpreneurs.

In the early nineteen-seventies in Gabon there were 18 sawmills producing annually about 50,000 m^3 of sawn wood and 20,000 m^3 of railway sleepers. More demands have been placed upon sawmilling in connection with changes in the production structure of the wood processing industry and as a consequence of the construction of the Transgabon Railway. The consumption of sawn wood and sleepers in the country increases. Therefore additional sawmills were established so as to increase sawn wood production to 150,000 m^3 and sleeper production to 280,000 m^3 by 1985. Yet this programme was not fulfiled.

Veneer sheets and plywood production is the most important branch of the wood processing industry in Gabon. At present annually about 90,000 m^3 of veneer sheets, mostly peeled from okoumé is produced. Plywood production is concentrated in Port Gentil, it reached a volume of about 80,000 m^3 at the beginning of the nineteen-seventies and was 130,000 m^3 in 1985. For plywood production mainly okoumé is used, and to a lesser degree ozigo, ilomba and other tree species are also processed.

Although the forest products industry lacks suitable technical equipment for production and has difficulty in competing with imported commodities of better quality, it has a relatively broad programme in furniture making and carpentry. Prefabricated wooden parts for housing are also manufactured for the local market.

Pulp production started in 1982 in the Kango plant near Libreville. The planned annual production of 175,000 tons of bleached sulphate cellulose in 1985 was not reached. Paper production is not envisaged so far.

Gabon is the second larger exporter of tropical timber from Africa (after the Ivory Coast) and also the largest exporter of okoumé logs. The share of okoumé in total log exports in 1961 was 90%. Following the year of the greatest boom in world commerce of tropical timber (1973) the export share of okoumé logs declined to 78%. Although the share of other tree species (ozigo, ilomba, tola, limba, acajou and douka) in log exports increased, the present situation does not correspond to the existing opportunities (Table 31). In stabilizing the export of okoumé logs an important role was played by the state institution the "Office des bois du Gabon" (renamed in 1976 as the "Societé nationale des bois du Gabon") which controls the felling and processing of timber, and supervises timber exports which are mainly of okoumé, ozigo and other tree species.

Table 31. Export of timber products from Gabon in 1000 m^3 (Yearbook of Forest Products, FAO, Rome, 1975, 1985)

Timber products	1965	1970	1973	1975	1980	1985
Logs	1,225	1,634	1,749	1,100	1,071	1,089
Sawn wood	22	15	5	2	18	1
Veneer sheets	20	21	14	9	19	8
Plywood	51	61	41	51	42	46

In spite of the effort of the governmental agencies in Gabon to promote a modern wood processing industry and to increase gradually the export of plywood, veneer sheets and other forest products, progress is slow. According to the Yearbook of Forest Products (FAO, 1981, 1986) the export value of timber and of the main forest products in Gabon represented 104.58 million US$ in 1973. From this total, logs represented 84% and plywood 11%. In 1985, 106.63 million US$ was the export value for timber and the main forest products in Gabon. In this sum, logs represented 86% and plywood 14%.

Gabon is a country with large resources of tropical timber and a small population. Even assuming a rising living standard of the population in the future, a large increase in local timber consumption cannot be expected. Thus it is obvious that Gabon will remain an exporter of tropical timber. It must be reckoned, of course, that ever more raw timber will be processed in the country and that the export of manufactured goods and semi-finished articles will be encouraged and the export of logs curbed.

In agreement with the official policy of non-alignment the Government of Gabon is interested in the expansion of economic cooperation in the field of logging and wood processing with foreign participation. This refers to logging and processing operations including enterprises for the manufacture of sawn wood, veneer sheets, plywood, and also for furniture making, joinery and carpentry, the manufacture of parquetry and packing material, charcoal production and the export of forest products in general.

4.4 Republic of Laos

4.4.1 Basic data

Laos is a land-locked country of an oblong form. It is situated in southeastern Asia between 13°15′ and 22°30′ N latitude and 97°10′ and 105° E longitude. It has a surface area of 236,800 km² and a population of about 4.3 million (1984), resulting in 18 inhabitants per square kilometer. Administratively Laos is divided into 18 provinces (Khouong) which are subdivided into districts (Muong) and further into communities (Tassong). The communities consist of villages and settlements (Ban). Vientiane, Savannakhet and Champassak are the most densely populated provinces. The capital of Laos is Vientiane. Lao is the official language. The monetary unit is the kip. In the map in Fig. 54 the numbers denote the following provinces: 1 – Phongsaly, 2 – Luang-Namtha, 3 – Oudomsay, 4 – Luang Prabang, 5 – Houa-Phanh, 6 – Sayaboury, 7 – Xieng-Khouang, 8 – Vientiane, 9 –Khammouane, 10 – Savannakhet, 11 – Saravane, 12 – Champassak, 13 –Attapeu.

Laos borders on Thailand, Burma, China, Vietnam and Cambodia. A large part of the border with Thailand is formed by the Mekong river. The natural border with Vietnam is formed by the Sai Phu Luong Mountains, also called the Annamese Cordillera.

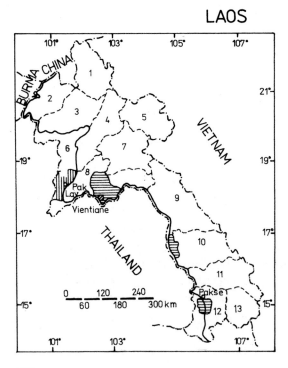

Fig. 54. Orientation map of the Republic of Laos.

More than 80% of the country's surface consists of plateaus and mountain massifs composed of granite, sandstone, slate and limestone. In northern and central Laos mountain peaks reach elevations of 2,000–2,800 m. The country's highest peak is Phou Bia (2,820 m above sea level). In southern Laos the highest mountain is Phou Set (1,540 m) on the Boloven uplands which is of volcanic origin. Towards the southwest and south the ground gradually descends to the valley of the Mekong river and its larger affluents (the Namhou, the Namkhan and the Nam Ngun).

About 60% of the country's surface is covered by forests, probably one third of them being commercial forests. On an area of about 5 million ha (26% of the forested area) devastated, unproductive forests occur which are the result of shifting agriculture, called the "ray" system. The most fertile agricultural land occurs in the lowlands at altitudes of 200 m along the Mekong river and other large watercourses.

The climate of Laos is tropical and moist with a pronounced dry and wet period. In the mountains the climate is subtropical. Annual precipitation varies between 1,300 mm (Luang Prabang) and 3,980 mm (Paksong). The wettest months are June, July and August when, for example, in southern Laos, in Paksan the average rainfall for these three months amounts to 1,800 mm. The dry period lasts from November to the middle of April. The average annual temperature is 26–27° C in the lowlands and 15–20° C in the mountains. In the hottest month of the year, namely in April, the average temperature attains 28–30° C in the lowlands (Table 32).

In Laos there are over 60 different ethnic groups. Officially three nationalities are recognized; the Lao Loum (Lao people of the lowlands). Lao Theoung (Kha people of the highlands), and Lao Soung (Meo-Lao people living in the mountains). The largest concentration of the population is in the Mekong valley.

Table 32. Climatic data from three stations of Laos

Climatic factors	Pakse	Vientiane	Luang Prabang
Northern latitude	15°07′	17°58′	19°53′
Altitude	96 m	160 m	350 m
Mean annual temperature	26.9°C	25.7°C	24.4°C
Mean annual maximal temperature	31.6°C	30.8°C	31.9°C
Mean annual minimal temperature	22.2°C	20.6°C	19.6°C
Mean annual rainfall	2,216 mm	1,715 mm	1,306 mm

Laos is one of the poorest developing countries in southeastern Asia. About 90% of the population depends on agriculture. Only 6% of the country's surface is devoted to permanent agriculture. Rice is the main agricultural crop, it is also cultivated on mountainous forest ground. Among other crops maize, potatoes, vegetables, tobacco, coffee, cotton and opium are important. Cattle are bred mainly as draught animals rather than as a source of meat, and are only exceptionally used for dairying. The Mekong river with its affluents is the basis for fisheries. With a view to the climate, the soil structure and water resources there are possibilities

125

for the faster growth of agricultural production, especially by extending the irrigation of rice fields, expanding vegetable plots and planting fruit trees.

Industry is underdeveloped and limited to woodworking which is mostly sawmilling, some food processing, small textile factories, a factory manufacturing corrugated iron sheets, and the production of medicaments and cigarettes. About 70% of the industrial installations, including a modern hydroelectric power station on the Nam Ngun river, are situated in Vientiane province. Eighty percent of the electrical energy produced by the above-mentioned power station is exported to Thailand.

As regards minerals, deposits of iron ore, zinc, lead, coal, copper, magnesite, silver and precious stones have been discovered. But with the exception of tin mines in Mong Sun and Phontiou no industrial plants for the exploitation of mineral resources have been set up so far.

Transportation is difficult because the road system is not sufficient. Most roads are not negotiable during the rainy season. River navigation is limited to a few navigable sections on the Mekong river. There are no railways in the country.

4.4.2 The types of forest formations and main economic tree species

From the point of view of the national economy of Laos forest resources play an overriding role. All forests are in state property. Not all forests are able to produce timber for industrial use.

According to a study published in 1973 by the Committee for the utilization of the Mekong river, the territory of Laos may be classified as follows (in 1,000 ha):

closed forest	5,380,
degraded closed forest	8,000,
open deciduous forest	1,940,
degraded open deciduous forest	310,
periodically inundated forest	165,
bush	5,235,
tree and bush vegetation on limestone	310,
savanna	850,
agricultural land	1,350,
water resources	140,
Total	23,680.

These statistical data do not indicate the area of land which is affected by shifting cultivation. It is estimated that at present about 100,000 ha of closed tropical forest is destroyed annually in Laos by shifting cultivation. Forest fires recurring in these regions almost regularly during the dry season cause the gradual destruction of the forest. A few years after deforestation the nutrients of the soil are exhausted, the land looses its fertility and shifting cultivation inevitably takes up another piece of forest land causing thereby irreparable damage to the ecosystem (Fig. 55).

126

Fig. 55. Devastated forests – shifting cultivation in the province of Vientiane, Laos.

In the work "La végétation du Laos" (Vidal, 1956) the author classified the forests of Laos into the following forest formations:

     1. Evergreen forests,

     2. Deciduous forests.

Evergreen forests are divided into:

     (a) dry evergreen forests of the lowlands,

     (b) dry evergreen forests of the hills,

     (c) evergreen forests of the mountains,

     (d) coniferous forests.

Deciduous forests are divided into:

     (a) dry dipterocarp forests – of high growth
                                 – of shrubby growth,

     (b) mixed deciduous forests.

1. *Dry evergreen forests* resemble tropical semi-deciduous forests by aspect and structure. They occur on various soil types up to an altitude of 1,000 m. They are the most important source of timber for sawn wood and fuelwood. These forest stands usually have two or more storeys, the largest trees growing to a diameter of 80–120 cm and to a height of 35–40 m. The trees which form an almost coherent canopy grow to a height of 20–30 m. Deciduous tree species with a proportion of up to 20% of the stand, shed their leaves in the dry period. The

127

main economic tree species belong to the *Dipterocarpus, Anisoptera* and *Hopea* genera.

1.(a) Dry evergreen forests of the lowlands grow up to an altitude of 200 m. They develop best on deep, well-drained alluvial soil. *Dipterocarpus alatus, Hopea ferrea* and, in the southern part of the country, *Parashorea stellata* are characteristic tree species of these lowland forests. In southern Laos *Dipterocarpus alatus* is less frequent. Because of the soil's fertility these lowland forests are often deforested to enable the cultivation of agricultural crops.

1.(b) Dry evergreen forests of the hills occur at altitudes above 200 m. Their aspects resembles that of the dry evergreen lowland forests, though the tree dimensions are here somewhat smaller. Here, *Dipterocarpus alatus* is less frequent and *Pterocarpus pedatus* and *Hopea ferrea* are characteristic economic tree species.
In both forest formations species of the *Lagerstroemia* genus and *Hopea odorata* occur frequently on somewhat drier and shallower soils.

1.(c) Mountain evergreen broad-leaved forests occur at altitudes over 1,000 m. The more important tree species of these forests are included in the *Quercus* and *Castanopsis* genera.

1.(d) Coniferous forests occur in the mountains and hills, and also on the lowlands, usually on poorer soil and steeper slopes. *Pinus merkusii* and *P. kesiya* are the main species. On higher elevations above sea level species of the *Quercus* and *Castanopsis* genera grow in the lower storey, on lowlands and hills species of the *Dipterocarpus* and *Pentacme* genera occur. Besides timber, resin and turpentines are also important products of coniferous forests.

2. *The deciduous tropical forests* of Laos consist mainly of dry Dipterocarp forests (Fig. 56). They grow on extensive areas in the lowlands and hilly country, mostly on less fertile and lateritic soils, or on steep dry slopes. The trees attain a height of 15–20 m, they shed their foliage at the beginning of the dry season.

2.(a) The characteristic tree species of Dipterocarp high forests are *Dipterocarpus tuberculatus, D. obtusifolius, D. intricatus, Pentacme siamensis* and *Terminalia tomentosa*. In shrubby dipterocarp forests the trees only reach a height of 10–15 m. The forests are open and grow on lateritic and gravely soils. Wood from these forests is used as fuelwood or for charcoal production. Species of the *Cratosylon* genus are characteristic tree species.

2.(b) In mixed deciduous forests of lowlands and hills *Afzelia xylocarpa, Xylia kerrii, Peltophorum dasyrachis, Pterocarpus macrocarpus* and species of the *Lagerstroemia* and *Vatica* genera are of frequent occurrence. About 85% of tree species shed their leaves in the dry period. In the central and nortwestern regions of Laos, teak – *Tectona grandis* – is an important tree species of rather moist deciduous mixed forests on hills.

Fig. 56. Dipterocarp forest. Stem of *Vatica dyerii* in the foreground. Phou pha Forest, Region of Vientiane, Laos.

On an area of about 600,000 ha bamboo forests grow, *Cephalostachyum, Oxytheutera* and *Dendrocalamus* genera being the main tree species. From an economic point of view, forests growing at altitudes of up to 1,000 m are of the greatest importance.

According to the publication "Forest Resources of Tropical Asia" (FAO, Rome, 1981) the area of forest land in Laos is 19,360,000 ha. Closed forests cover 8,410,000 ha (43%), open savanna-like forests (woodland savannas) cover 5,215,000 ha (27%); unproductive forest land devastated by shifting cultivation is found on an area of about 5 million ha (26%), and bush covers about 735,000 ha (4%) of forest land.

Forest tree species in Laos are classified according to their utilization into the following four categories (McCombe and Krieg, 1977):

(A) Luxury tree species:
   Afzelia xylocarpa CRAIB. – mai kha, merbau
   Cassia siamea LAM. – m. khi lek
   Dalbergia bariensis PIERRE – m. kham phi
   D. cochinchinensis PIERRE – m. kha nhoung
   Diospyros mollis GRIFF. – m. lang dam
   D. mun H. LEC. – m. nang dam
   Dysoxylum spp. BLUME – m. khon tasang
   Fagraea fragrans ROXB. – m. manh pa
   Melanorrhoea laccifera PIERRE – m. nam kieng
   Pterocarpus macrocarpus KURZ – m. dou
   P. pedatus PIERRE – m. dou
   Tectona grandis L. f. – m. sat; teak
   Cunninghamia sinensis R. BR. – m. long leng
   Fokienia kawai – m. len le
(B) Tree species of the first category:
   Artocarpus integrifolia L. – mai mi
   Chukrasia tabularis A. JUSS. – m. nhom hin
   Dialium cochinchinensis PIERRE – m. kheng; sepan
   Dipterocarpus alatus ROXB. – m. nhang; yang; keruing
   Garcinia fragraeoides A. CHEV. – m. trai ly
   Hopea ferrea PIERRE – m. khen hin; balau
   H. odorata ROXB. – m. khen hua; merawan; sao
   H. pierrei HANCE – m. laen; merawan; sao
   Lagerstroemia spp. L. – m. puay
   Litsea polyantha JUSS. – m. mee
   Mesua ferrea L. – m. khathang; bunnark
   Shorea cochinchinensis BAILL. – m. kha nhom; white meranti
   S. obtusa WALL. – m. chick
   Sindora cochinchinensis BAILL. – m. te; sindoer
   Stereospermum spp. CHAM. – m. khe foy
   Talauma gioi A. CHEV. – m. ham; champak
   Vatica dyeri PIERRE – m. si deng; resak
   V. roxburghiana BLUME – m. chick dong; resak
   Xylia xylocarpa TAUB. – m. deng; pyinkado
   Ketelleria davidiana BEISSN. – m. hing
(C) Tree species of the second category:
   Aglaia gigantea PELLEGR. – mai nok kok; tasua
   Anisoptera cochinchinensis PIERRE – m. bak; krabak, mersawa
   Castanopsis spp. SPACH. – m. ko nam
   Dipterocarpus intricatus DYER. – m. sa beng
   D. obtusifolius TEYSM. – m. sat; yang
   D. tuberculatus ROXB. – m. koung

Heritiera javanica KOST. – m. hao
Manglietia glauca BLUME – m. luong khom; champak
Pentacme siamensis KURZ – m. hang
Pygeum arboreum ENGL. – m. mak tek
Quercus spp. L. – m. ko
Shorea floribunda WALL. – m. kha nohom
S. hypochra HANCE – m. khen khay; white meranti
S. vulgaris PIERRE – m. si khao
Terminalia tomentosa BEDD. – m. suak; Indian laurel
Toona febrifuga ROEM. – m. nhom hom; suren
Pinus kesiya ROYLE ex GORDON – m. pek
P. merkusii JUNGH et DE VRIESE – m. pek

(D) Tree species of the third category:
Adina cordifolia HOOK. f. – mai khao; haldu
Canarium nigrum ENGL. – m. bay; kedondong
Ficus spp. L. – m. dua
Machilus spp. NEES – m. khen dong
Melia spp. L. – m. hien
Pterospermum spp. SCHREBER – m. hang hen
Sandoricum indicum CAV. – m. tong; kra-thon
and other tree species.

In the above-mentioned list there is a total of 60 tree species which in Laos are of economic importance either for export or for local utilization. There is a large number of tree species in the forests of Laos, estimated at 200–250, which could be used in the national economy. From among this latter group a few species of some economic importance which occur more or less frequently in certain localities in Laos are mentioned below:

Ailanthus malabarica DC. – mai nhom pha
Anisoptera costata KORTH – m. bak; krabak
Carallia lucida ROXB. – m. bong nang
Cinnamomum inners WALL. – m. si khai
Dalbergia cultrata GRAH. – m. kham phi
D. oliveri GAMB. – mai padong
D. lanceolaria L. – m. kham phi
(all these three tree species are included among Asian palisanders)
Dipterocarpus costatus GAERTN. f. – m. nhang deng
D. turbinatus GAERTN. f. – m. nhang
Irvingia harmandiana – m. bok; kabok
Parashorea stellata KURZ – m. hao
Shorea talura ROXB. – m. kha nhom; white meranti
Terminalia chebula RETZ. – m. som mo; chuglam
Vitex pubescens VAHL. – m. tin nok; thom
Xylia kerrii CRAIB. et HUTCHINS. – m. deng; pyinkado.

Fig. 57. Stem of *Dipterocarpus alatus*. D.B.H. = 95 cm. Phou Pha Forest, Laos.

A concise characteristics, properties and possibilities of use for some economically important tree species are given below:

Afzelia xylocarpa CRAIB.

(Syn.: *Pahudia cochinchinensis* PIERRE; *Afzelia cochinchinensis* J. LÉONARD)

Laotian name: mai kha. Commercial names: merbau, beng.

A tree with deciduous leaves, growing to a height of 30 m and a diameter of 100–130 cm. Heartwood of light brown to red colour with a fine texture resembling mahogany timber. The wood is heavy, specific weight: 0.80 to 0.90, hard, easy to process and polish. It is used for luxury furniture, veneer sheets and plywood production, for parquetry and in construction.

Anisoptera cochinchinensis PIERRE

Laotian name: mai bak. Commercial names: krabak, mersawa.

An evergreen tree growing according to localities to a height of 25–45 m and to a diameter of 80–130 cm. The wood is of yellow to yellowish-brown colour, turning dark in the course of time, and containing silicates. It is medium heavy, the specific weight is 0.60–0.72, not very durable, but easy to work. The sapwood differs little from the heartwood, and is attacked rapidly by mould and insects.

Anisoptera cochinchinensis is the second most frequently exploited tree species in Laos. It is used for sliced and peeled veneer sheets, for flooring, furniture and in carpentry.

Dalbergia cochinchinensis PIERRE

Laotian name: mai kha nhoung. Commercial names: Asian palisander, Siam rosewood.

The tree, with deciduous leaves, grows to a height of 25–30 m and to a diameter of 80 cm. Its stem is a straight, cylindrical bole. It occurs singly or in clusters in both dry evergreen and mixed deciduous forests.

The timber is of mauve colour, with dark stains, hard and very heavy; the average specific weight is 1.08. It is durable, easy to work and exceedingly good for polishing. It is used for wood-carving, massive luxury furniture, decorative veneer sheets and in the cabinet-maker's trade.

Similar properties are found also in the timber of *Dalbergia bariensis* PIERRE, mai kham phi, which is of a wine-red colour; and also in *Dalbergia oliveri* GAMBLE, mai padong, which has a reddish-brown colour with dark stains. The timber of *Dalbergia cultrata* GRAH., mai kham phi, is very decorative, durable, hard and heavy and of a copper-red to black colour, it is difficult to work.

Diospyros mun H. LEC.

Laotian name: mai nang dam.

A tree growing to the height of 20 m and to a diameter of 60–80 cm. The timber is of dark green colour with a homogeneous fine-grained structure, decorative, durable, very hard and very heavy, the specific weight is 1.30. It is easy to work and polish. It is used in wood-carving, the cabinet-maker's trade and in the manufacture of musical instruments.

Ebonies are also economically important as follows:

Heritiera javanica KOST. – m. hao

Manglietia glauca BLUME – m. luong khom; champak

Pentacme siamensis KURZ – m. hang

Pygeum arboreum ENGL. – m. mak tek

Quercus spp. L. – m. ko

Shorea floribunda WALL. – m. kha nohom

S. hypochra HANCE – m. khen khay; white meranti

S. vulgaris PIERRE – m. si khao

Terminalia tomentosa BEDD. – m. suak; Indian laurel

Toona febrifuga ROEM. – m. nhom hom; suren

Pinus kesiya ROYLE ex GORDON – m. pek

P. merkusii JUNGH et DE VRIESE – m. pek

(D) Tree species of the third category:

Adina cordifolia HOOK. f. – mai khao; haldu

Canarium nigrum ENGL. – m. bay; kedondong

Ficus spp. L. – m. dua

Machilus spp. NEES – m. khen dong

Melia spp. L. – m. hien

Pterospermum spp. SCHREBER – m. hang hen

Sandoricum indicum CAV. – m. tong; kra-thon

and other tree species.

   In the above-mentioned list there is a total of 60 tree species which in Laos are of economic importance either for export or for local utilization. There is a large number of tree species in the forests of Laos, estimated at 200–250, which could be used in the national economy. From among this latter group a few species of some economic importance which occur more or less frequently in certain localities in Laos are mentioned below:

Ailanthus malabarica DC. – mai nhom pha

Anisoptera costata KORTH – m. bak; krabak

Carallia lucida ROXB. – m. bong nang

Cinnamomum inners WALL. – m. si khai

Dalbergia cultrata GRAH. – m. kham phi

D. oliveri GAMB. – mai padong

D. lanceolaria L. – m. kham phi

(all these three tree species are included among Asian palisanders)

Dipterocarpus costatus GAERTN. f. – m. nhang deng

D. turbinatus GAERTN. f. – m. nhang

Irvingia harmandiana – m. bok; kabok

Parashorea stellata KURZ – m. hao

Shorea talura ROXB. – m. kha nhom; white meranti

Terminalia chebula RETZ. – m. som mo; chuglam

Vitex pubescens VAHL. – m. tin nok; thom

Xylia kerrii CRAIB. et HUTCHINS. – m. deng; pyinkado.

is hard and heavy, with a specific weight of 0.75–0.90, durable, resistant to insect attack and injury by termites and fungi, and it is also weather resistant. It is used for luxury furniture, wood-carving, decorative veneer sheets, flooring and in construction.

Shorea hypochra HANCE
(Syn.: *Shorea crassifolia* RIDL.)
Laotian name: mai khen khay. Commercial name: white meranti.

This tree grows to a height of 40 m and to a diameter of 100 cm. The yellow timber resembles satinwood *(Chloroxylon swietenia* A.DC.). The structure of the timber is finely grained and homogeneous. The timber is medium hard and medium heavy, its specific weight is 0.70–0.75. It is attacked by termites. It is used for internal construction but mainly for plywood production.

Shorea vulgaris P. (in Laotian: mai si khao) is also of rather frequent occurrence in Laos.

This is a 30–40 m high tree with a diameter of 120 cm. It occurs in dry evergreen, deciduous mixed and Dipterocarp forests. The timber is hard and heavy, medium durable but liable to blue stain. It is used for construction, carpentry, flooring, railway sleepers and plywood.

Shorea talura ROXB.
(Syn.: *Shorea harmandii* LANESS.).
Laotian name: mai kha nhom. Commercial name: red meranti.

This tree attains a height of 20–30 m and a diameter of 120 cm; the stems are frequently crooked. The timber is yellow-brown to red-brown in colour, it is hard and heavy, the specific weight is 0.82–0.93. It is medium durable, of intermediate workability and exceedingly good for polishing. It is used in construction, for furniture and decorative veneer sheets.

Sindora cochinchinensis BAILL.
Laotian name: mai te. Commercial name: sindoer.

This is a tree with deciduous leaves growing to a height of 35 m and to a diameter of 100–130 cm.

The heartwood is of a yellow-brown to russet colour. The structure of the timber is fine, with parallel fibres, sometimes showing a strip design. The timber is medium hard, medium heavy, with a specific weight of 0.72–0.83, durable, resistant to insects and termites. It is easy to work, slice and peel, and is exceedingly good for polishing. It is used for decorative veneer sheets, plywood, luxury furniture and in wood-carving.

Xylia kerrii CRAIB et HUTCH.
Laotian name: mai deng. Commercial names: pyinkado, resak.

This is a tree with deciduous leaves growing to a height of 30 m and to a diameter of 100–150 cm. It forms straight cylindrical boles.

The red-brown heartwood turns dark as time goes by. The timber has twisting fibres, a fine design, it is decorative, very strong, hard and heavy; its specific weight is 0.90–0.98. It is resistant to insects, termites and fungi. It is rather difficult to work and is not suitable for slicing and peeling. It is used in heavy construction, bridge building, flooring, parquetry and for railway sleepers.

4.4.3 Results of forest inventories in certain localities of Laos

For the first time a forest inventory in Laos was made during 1967–1969. The Canadian Agency for International Development (CAID) (Paquet et al., 1969) inventoried forests in three localities near to the Mekong Valley covering 347,691 ha of forests in the Vientiane province, 141,102 ha of forests in the Savannakhet province and 127,190 ha of forests in the Pakse province; making a total of 615,983 ha of forests surveyed. The area of non-forested land in the inventoried region was 262,881 ha, i.e. 30% of the total area of the inventoried region (878,864 ha).

The aim of the inventory work was to obtain basic information on the structure of the inventoried forests on selected localities; such data being necessary for the development of forestry and forest industries in Laos. Information on empirical data from this inventory was obtained using the archives of the forestry headquarters. Proper inventory work was preceded by aerial photography of the region. After the evaluation of the aerial photographs a preliminary reconnaissance of the ground was made and on this basis a systematic network of inventory sample plots was established.

In the Savannakhet region a simple procedure was chosen. The inventory plots were demarcated, keeping in line with the aerial photographs, where the greatest variability in species composition and growing stock was expected. In one statistical unit a total of ten circular inventory plots was included.

In the Vientiane and Pakse regions a more complicated procedure in marking out the inventory plots on the ground – the camp system – was chosen. In accordance with the aerial photographs, the inventory plots were concentrated in localities with an area of 25 km² of forest land. In the centre of the inventoried locality the "camp" of the inventory team was established. The circular inventory sample plots were grouped at a distance of 80 m from their centres into statistical units – clusters – forming a rectangle of 240×160 m (Fig. 58) (this was done for practical reasons, too, in order to shorten the walking distance). Fourty such statistical units were systematically distributed at a distance of 500 m on both sides of two ideal lines which were parallel to each other, 5 km long, and which usually intersected in the centre of the camp. Thus one camp included 40 statistical units. The daily norm of the inventory group was to inventory one statistical unit, or 10 circular inventory plots. The circular inventory plots were divided into four smaller concentric areas where an inventory was carried out according to the pattern given in Table 33.

During inventory work the situation of the trees was recorded, the tree species was identified and the diameter breast height was measured. For trees with a diame-

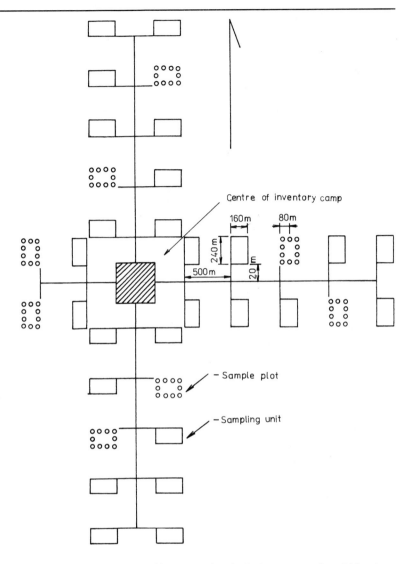

Fig. 58. Scheme of the tracing of inventory plots in the inventory region of Vientiane and Pakse, Laos.

ter of up to 30 cm the height of the stem was measured to the first branches, in trees with a diameter above 30 cm the height and the diameter of the stem were measured in sections of 5 m.

From the quality point of view the trees were classified into the following classes:

First class – sound trees with straight cylindrical boles,

Second class – sound trees but crooked and having other defects,

Third class – dying trees from which at least 50% of the stem volume can be salvaged.

Table 33. Scheme of sampling in circular sample plots

Radius	Area	Minimum diameter of trees	Category of the stand
of the plot			
m	ha	cm	
1.26	0.005	1_+	Thicket
5.64	0.01	10_+	Pole stage
5.65	0.01	5_+	Bambou forest
		(bamboos over 4 m)	
17.84	0.10	30_+	Small sized woods
28.21	0.25	60_+	Stem timber crop

In computing the standing timber stock the formulae were used which were applied for computing the growing stock in the forest inventory carried out in neighbouring Cambodia by USAID.

In deciduous tropical forests for all tree species with a diameter breast height of up to 30 cm the formula:

$$V = 0.02756 + 1.49511D^2H \text{ was used,}$$

and for all trees with a diameter breast height above 30 cm the formula:

$$V = 0.00156 + 1.44890D^2 + 1.40889D^2H$$

was applied.

In evergreen forests likewise for all tree species with a diameter breast height below 30 cm one formula has been used, namely

$$V = 0.02197 + 1.67851D^2H,$$

and for trees with a diameter exceeding 30 cm the formula:

$$V = 0.28053 + 1.89533D^2H$$

was employed.

In the above-mentioned formulae the following designation is used: V is the stem volume, D the diameter breast height (D.B.H.) of stems with bark, H the stem length (to the first branches).

The representation of the main forest formations in both absolute and relative units in the inventoried regions is given in Table 34.

The inventory results showed that in the Vientiane region tropical lowland mixed deciduous forests occur most frequently, representing 42% of all forests in the inventoried region.

Table 34. Tropical forest formations in inventoried regions of Laos

Type of forest formation	Vientiane		Savannakhet		Pakse	
	ha	%	ha	%	ha	%
Lowland mixed deciduous forests	146,775	42.2	48,095	34.1	52.623	41.4
Lowland dry evergreen forests	44,334	12.8	14,647	10.4	25,607	20.1
Dry evergreen hill forests	22,262	6.4	–	–	28,845	22.7
Dry deciduous diptero-carp forests	24,523	7.0	77,994	55.3	6,046	4.8
Mixed deciduous hill forests	87,913	25.3	–	–	7,878	6.2
Bamboo forests	21,884	6.3	366	0.2	6,191	4.8
Total	347,691	100.0	141,102	100.0	127,190	100.0

In the tropical lowland mixed deciduous forests species of the *Lagerstroemia* genus with 29 trees per hectare of a diameter exceeding 10 cm, and species of the *Diospyros* genus with 9 trees per hectare were the most represented.

Species of the *Diospyros* genus were found growing only to a diameter of 60 cm. The average number of trees per hectare (computed from 442 plots each one measuring 0.25 ha) with a diameter above 10 cm was 238, and 11 trees were found on average with a diameter above 60 cm.

The average growing stock per hectare for trees with a diameter breast height greater than 10 cm was 93 m^3 and for trees with a diameter above 60 cm it was 18 m^3.

The lowland evergreen forests are the most productive of the forest formations in the Vientiane region but they represent only 13% of the forests in the Vientiane inventory region. This formation occurs mainly on rich soils.

In the evergreen dry forests *Hopea ferrea* (16 trees per hectare with a diameter above 10 cm), species of the *Diospyros* genus (13 trees per hectare) and species of the *Lagerstroemia* genus (12 trees per hectare) are the most represented. The average number of trees per hectare computed from 262 plots (each one measuring 0.25 ha) was 228, and for trees with a diameter breast height above 60 cm it was 13 trees per hectare. The average growing stock for trees with a diameter breast height greater than 10 cm was 167 m^3 per hectare, and for trees with a diameter above 60 cm it was 66 m^3 per hectare.

The so-called bamboo forests are the least productive forests in the inventoried regions. They are a degraded stage of the forest which represents in the Vientiane inventory region over 6% of all forests. From among commercial tree species *Sandoricum indicum* (10 trees with a diameter more than 10 cm per hectare) and species of the *Diospyros* genus (7 trees per hectare) were the most represented. The average number of trees per hectare including bamboo thickets (92 pieces) was 227 and for trees with a diameter larger than 60 cm it was only one tree. These data were obtained from 142 inventory plots each with an area of 0.25 ha. The average growing stock for trees with a diameter greater than 10 cm was only 31 m³ per hectare and for trees with a diameter of more than 60 cm it was 3 m³ per hectare.

In the Savannakhet inventory region dry dipterocarp forests prevail which cover 55% of the area of all forests in the inventoried region. Trees with a diameter breast height over 100 cm were not identified. Among trees with a diameter breast height greater than 10 cm *Shorea obtusa* (75 trees per hectare) and *Dipterocarpus tuberculatus* (48 trees per hectare) were the most common species. The average number of trees per hectare with a diameter more than 10 cm, computed from 435 plots with an area of 0.25 ha, was 266, but only 2 trees per hectare for a tree diameter exceeding 60 cm. The average growing stock for trees with a diameter above 10 cm was 55 m³ per hectare, and 3 m³ per hectare for trees with a diameter more than 60 cm.

Lowland mixed deciduous forests are the second largest forest formation in the Savannakhet region representing 34% of the area of all forests in the inventoried region. From among commercial tree species *Peltophorum dasyrachis* (28 trees per hectare with a diameter greater than 10 cm) and species of the *Lagerstroemia* genus (24 trees per hectare) are the most represented in these lowland mixed forests. The average number of trees per hectare with a diameter above 10 cm (computed from 279 inventory plots having an area of 0.25 ha) was 315, and for trees with a diameter above 60 cm it was 10 trees per hectare. The average growing stock of trees with a diameter above 10 cm was 107 m³ per hectare and for trees with a diameter above 60 cm it was 7 m³ per hectare.

The lowland evergreen forests are also in the Savannakhet region the most productive forests and they represent 10% of the area of all forests in the inventoried region. Species of the *Diospyros* genus (39 trees per hectare with a diameter more than 10 cm) and *Shorea vulgaris* (18 trees per hectare) were the most common commercial tree species in these forests. The average number of trees per hectare (computed from 272 inventory plots having an area of 0.25 ha) was 305, and for trees with a diameter above 60 cm it was 12 trees per hectare. The average growing stock of trees with a diameter above 10 cm was 165 m³ per hectare, and for trees with a diameter above 60 cm it was 43 m³ per hectare. The largest amount of mature timber resources (D.B.H. ≥ 60 cm) belonged to *Dipterocarpus alatus* (8 m³ per hectare), *Anisoptera cochinchinensis* and *Dipterocarpus tuberculatus* (each 7 m³ per hectare) and *Canarium nigrum* (4 m³ per hectare).

In southern Laos, in the Pakse inventory region, lowland mixed deciduous forests are most widespread and cover 41% of the area of all forests in the inventoried region. In these mixed forests species of the *Diospyros* genus (39 trees per hectare) and of the *Lagerstroemia* genus (18 trees per hectare with a diameter above 10 cm) were the most common commercial tree species. Relatively large diameter dimensions are mainly found in species of the *Lagerstroemia* genus and in *Dipterocarpus tuberculatus*. The average number of trees per hectare computed from 64 inventory plots (having an area of 0.25 ha each) was 329 for trees with a diameter of 10 cm and more, and 8 trees per hectare for trees with a diameter above 60 cm. The average growing stock was 120 m³ per hectare for trees with a diameter exceeding 10 cm, and 15 m³ per hectare for trees with a diameter exceeding 60 cm.

The mixed deciduous forests of the hills cover 6% of the forest area in the Pakse inventory region. Because inventory work here was confined to only 13 sample plots with a total area of 3.25 ha, the results may not give an adequate picture of the situation. In the mixed deciduous forests of this region *Vitex pubescens* (58 trees per hectare with a diameter above 10 cm), species of the *Diospyros* genus (48 trees per hectare with a diameter of only up to 45 cm) and species of the *Lagerstroemia* genus (45 trees per hectare) were most common. The average number of trees per hectare (computed from 13 inventory plots having an area of 0.25 ha each), was 428 for trees with a diameter exceeding 10 cm, and 12 for trees with a diameter exceeding 60 cm. The average growing stock was 118 m³ per hectare for trees with a diameter exceeding 10 cm and 19 m³ per hectare for trees with a diameter exceeding 60 cm.

In the Pakse region evergreen forests are frequent, covering 43% of the area of all forests in the inventory region; they include 20% of lowland evergreen forests and 23% of evergreen forests on the hills. In the lowland evergreen forests species of the *Diospyros* genus (42 trees per hectare starting with a diameter of 10 cm) and *Dipterocarpus tuberculatus* (14 trees per hectare) were the most common tree species. The average number of trees per hectare, computed from 88 inventory plots (each one measuring 0.25 ha), was 363 for trees with a diameter exceeding 10 cm and 9 for trees with a diameter exceeding 60 cm. The average growing stock was 178 m³ per hectare for trees with a diameter greater than 10 cm, and 30 m³ per hectare for trees with a diameter above 60 cm.

In evergreen forests of the hills in the Pakse region species of the *Diospyros* genus were also very frequent, though less common than in the evergreen forests of the lowlands (24 trees per hectare for diameters above 10 cm). Well represented species were *Hopea ferrea* (23 trees per hectare) and *Vatica dyeri* (12 trees per hectare). The average number of trees per hectare, computed from 54 inventory plots for the diameters above 10 cm was 371, and for tree diameters above 60 cm it was 14 trees per hectare. The average growing stock was 172 m³ per hectare for trees with a diameter above 10 cm, and 43 m³ per hectare for trees with a diameter above 60 cm.

In the Laotian forests many tree species grow to a diameter breast height of 40–60 cm. These include, for example, certain species of the *Dalbergia*, *Diospyros* and *Melia* genera, as well as *Melanorrhoea laccifera*. Trees with a diameter breast height exceeding 100 cm are found for the following tree species: *Afzelia xylocarpa*, *Anisoptera cochinchinensis*, species of the *Bailschmiedia* genus, *Canarium nigrum*, *Chukrasia tabularis*, *Dialium cochinchinensis*, *Dipterocarpus alatus*, *D. intricatus*, *D. tuberculatus*, *Hopea ferrea*, *H. odorata*, species of the *Lagerstroemia* genus, *Sandoricum indicum*, *Shorea vulgaris*, *Sindora cochinchinensis*, species of the *Machilus* genus, *Pentacme siamensis* and *Pterocarpus macrocarpus*.

Sporadically, some tree species grow to large dimensions. Trees with a diameter breast height exceeding 200 cm were identified during inventory work in Laos for the following species: *Anisoptera cochinchinensis, Dipterocarpus alatus, Heritiera javanica* and *Irvingia malayana.*

Taking into consideration the growth capacity of tree species, official permission has been given to fell certain species such as *Dalbergia bariensis, D. cochinchinensis, Diospyros mollis, Shorea obtusa,* and also *Tectona grandis* where these have a D.B.H. greater than 30 cm. Other species may be cut starting from a diameter of 40 cm, these include *Afzelia xylocarpa, Dalbergia cultrata, Chukrasia tabularis, Lagerstroemia floribunda, Hopea odorata, Pterocarpus macrocarpus, Vatica dyeri* and *Xylia kerrii.* For the felling of other species the minimum cutting diameter has been established at 45 cm.

The distribution of trees by diameter breast heights with regard to single tree species or to their groups has been expressed using the well-proved mathematical formula for the beta function. Apart from sporadic instances, the use of the beta function for the compensation of the declining tendency in the distribution of the number of trees according to their diameter, proved to be convenient in the natural tropical forests of Laos.

Using inventory data from field notes the values of the diameter distribution of trees were mathematically compensated, according to the beta function, summarily for the main forest formations in each inventory region. The compensated values derived thus (converted to an area of one hectare), are given in Table 35 for the Vientiane region and in Table 36 for the Pakse region.

Compensated data of the diameter distribution of trees for the Pakse inventory region in the south of Laos are shown in Fig. 59. All compensated curves showing the tree distribution by diameter breast height in the various forest formations have a clearly declining tendency with increasing tree diameter.

The dry deciduous Dipterocarp forests in the Vientiane and Savannakhet regions are marked by the largest representation of trees in classes of low diameters. Larger dimensions of trees, exceeding 90 cm, occur only seldomly in this forest formation.

Lowland evergreen forests, as to the creation of growing stock, are the most productive formations in all three inventory localities. The curves of tree frequency show a more moderate declining tendency from the smallest to the largest diameter classes.

Numerical data of the diameter distribution of trees for *Hopea ferrea* from four different localities are mathematically compensated according to the beta function in Table 37. This important economic tree species is of frequent occurrence and grows to large dimensions on all observed localities.

Taking into account that 45 cm is the minimum cutting diameter allowed for *Hopea ferrea,* the figures of the diameter distribution of trees clearly indicate important resources of mature timber of this commercial tree species.

In 1980 a group of Czechoslovak foresters in Laos carried out an inventory survey

Table 35. Theoretical diameter breast height distribution of trees of the main forest formations by the beta-function. Inventory region Vientiane, Laos

Midpoint values of D.B.H. classes	Tropical forest formations			
	Dry evergreen forests		Dry dipterocarp deciduous forests	Mixed deciduous forests of lowlands
	lowlands	hills		
cm	Number of trees per ha			
25	56	47	75	50
35	24	22	35	23
45	14	13	17	14
55	10	9	9	9
65	7	6	4	6
75	5	4	2	4
85	3	2	1	3
95	2	2	–	2
105+	1	1	–	1
Total	122	106	143	112
Const.	0,0083	0,0010	0,0000	0,0034
α	−0,6086	−0,5134	−0,2477	−0,5520
β	1,5718	1,9515	5,2162	1,7221
χ^2	16,425	7,948	7,090	12,156

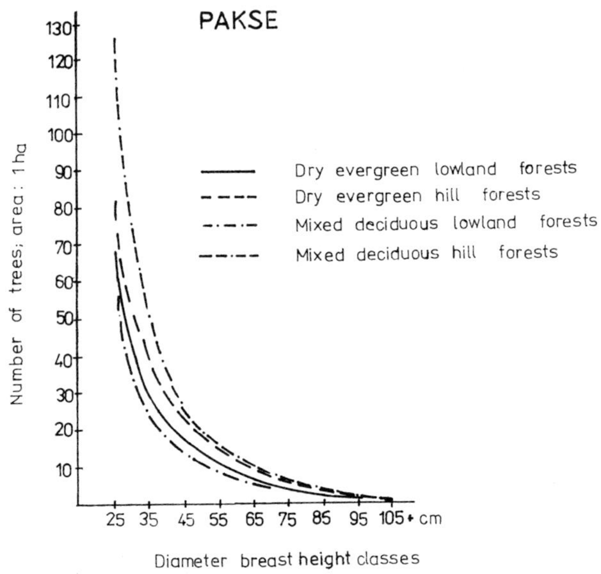

Fig. 59. Compensated values of the diameter distribution of trees according to the beta function in the main forest formations. Region of Pakse, Laos.

PAKSE

Number of trees; area: 1 ha

Dry evergreen lowland forests
Dry evergreen hill forests
Mixed deciduous lowland forests
Mixed deciduous hill forests

Diameter breast height classes

Table 36. Theoretical data of the beta-diameter distribution of trees of main forest formations. Inventoried region Pakse, Laos

Midpoint values of D.B.H. classes	Tropical forest formations			
	Dry evergreen forests		Mixed deciduous forests	
	lowlands	hills	lowlands	hills
cm	Number of trees per ha			
25	67	81	57	124
35	29	38	24	51
45	15	22	13	25
55	10	14	8	15
65	6	9	6	9
75	4	5	3	5
85	2	3	2	3
95	1	2	1	1
105+	1	1	1	1
Total	135	175	115	234
Const.	0.0003	0.0001	0.0004	0.0001
α	−0.5799	−0.4356	−0.5952	−0.6068
β	2.3436	2.6765	2.1921	2.7536
χ^2	17.929	8.979	16.080	37.234

around the town of Paklay in the Sayabouri province (17°50′ to 18°55′ N latitude and 100° and 100°30′ E longitude).

The inventory region borders, in the east on the Mekong river, in the south on the Nam Houng river, in the north on the Nam Pouy watercourse and in the west on Thailand. The area of the inventory region was about 590,000 ha. The forest area amounted to about 330,000 ha and included 20% of closed productive forests, 60% of semi-deciduous and deciduous degraded forests and about 20% of deforested land and land with rice fields and other agricultural crops (Černý et al., 1980).

In the Paklay inventory region the ground is very rough and broken and is situated at an altitude from 200 to 1,800 m. Average annual precipitation is around 1,400 mm. The average monthly temperature varies between 20 and 28°C. The minimum recorded temperature was 3°C, the maximum was 45°C.

All the trees with a diameter breast height of more than 20 cm were measured within circular inventory plots which had a total area of 40 ha of forests. The area of the circular inventory plots was 0.25 ha.

On strip inventory plots, each with an area of 2 ha (20 m width), only trees with a larger diameter (D.B.H. ≥ 50 cm) were callipered. Luxury tree species such as

144

Table 37. Theoretical diameter breast height distribution of *Hopea ferrea* trees in main tropical forest formations by the beta function, Laos

Midpoint values of D.B.H. classes	Mixed deciduous forests	Dry evergreen forests		
		lowlands	hills	
	Pakse	Vientiane	Vientiane	Pakse
cm	Number of trees per 10 ha			
25	63	51	72	65
35	26	29	50	37
45	12	18	36	25
55	6	11	26	18
65	3	6	18	13
75	1	3	11	8
85	1	1	7	5
95	–	1	4	3
105	–	–	2	–
Total	112	120	226	174
Const.	0.0000	0.0000	0.0000	0.0051
α	−0.4386	−0.1703	−0.0876	−0.3176
β	3.4375	3.2590	2.6936	1,6697
χ^2	9.356	2.261	3.407	3.353

species of the *Dalbergia* and *Diospyros* genera were callipered starting from a diameter breast height of 20 cm. The total area of strip inventory plots was about 140 ha.

On the basis of the inventory, it may be stated that in the Paklay region of Sayabouri province the following main forest formation occur (Černý et al., 1980):

1. dry evergreen tropical forests,
2. mixed deciduous tropical forests,
3. dry, deciduous open dipterocarp forests.

Among commercial tree species in the dry evergreen tropical forests of the lowlands and the hills *Lagerstroemia speciosa* and *L. calyculata* are the most common; with 15 trees per hectare having a diameter breast height above 20 cm, and 8 trees per hectare with a diameter breast height over 50 cm. Both tree species occurred in larger dimensions up to a diameter breast height of 150 cm and to a height of 25 to 30 m; they formed knotty stems of bad shape.

From among other commercial tree species *Dipterocarpus alatus* with 2–5 trees per hectare, *Afzelia xylocarpa* with 1 to 3 trees per hectare and *Sindora cochinchinensis* with 1–2 trees per hectare occurred more frequently, all these trees having a diameter breast height greater than 50 cm. Species of the *Diospyros* genus occurred on average with 5 trees per hectare, having a diameter above 20 cm. From among lesser known tree species in the formation of dry evergreen forests, species of the *Polyalthia* genus (with 20 trees per hectare of a diameter breast height above 20 cm) and *Ormosia cambodiana, Pometia eximia, Ficus annulata* and *Irvingia harmandiana* were the most represented. These trees at present have no economic importance although they are used locally for fuelwood.

The mixed semi-deciduous and deciduous tropical forests in the Paklay region are characterized by a larger occurrence of commercial tree species. The main one is *Dipterocarpus obtusifolius* which is represented by 30 trees per hectare with a diameter exceeding 20 cm, and by 10 trees per hectare with a diameter above 50 cm. In addition *Dipterocarpus alatus* was represented by 8–11 trees per hectare with a diameter above 20 cm, and by 3 to 4 trees per hectare, with a diameter above 50 cm.

One of the most important commercial tree species occurring in this forest formation is *Tectona grandis*, teak. In the Paklay region it grows to a diameter of 100 cm and to a height of 20 m. On inventory sample plots with a total area of 11 ha, 237 trees with a diameter larger than 20 cm were recorded, i.e. 21 trees per hectare (in-

Table 38. Theoretical data of the beta-diameter distribution of trees of some timber species. Inventoried region Paklay

Midpoint values of D.B.H. classes	*Dipterocarpus alatus*	*Pterocarpus macrocarpus*	*Tectona grandis*	*Dalbergia cochinchinensis*
	Mixed deciduous forests			Dipterocarp forests
cm	Number of trees per ha			
25	30	27	54	22
35	25	16	58	15
45	19	11	45	10
55	13	7	30	6
65	9	4	18	3
75	6	3	10	–
85	3	1	1	–
95	1	1	1	–
105	1	1	–	–
Total	107	71	217	56
Const.	0.0004	0.0000	0.0000	0.0643
α	0.0227	−0.1531	0.4224	−0.1960
β	2.0191	3.0301	2.7963	0.9781
χ^2	8.408	4.704	22.842	0.480

cluding 6 trees per hectare with a diameter breast height above 50 cm). The timber of teak is very durable, it is of universal use, decorative and very valuable.

As to timber production in the Paklay region dry tropical dipterocarp forests (also called in French literature "forêt claire") are the poorest formation. They represent open forest stands with an under storey of shrubs and grasses where tree species grow to a diameter of 50–60 cm, exceptionally 80–100 cm. Such species include *Ficus annulata, Afzelia xylocarpa, Dipterocarpus alatus, Pentacme siamensis* and species of the *Lagerstroemia* genus. Trees in the upper storey reach a height of 18–22 m.

In the Paklay inventory region *Pentacme siamensis* was the most common commercial species, numbering 18 trees per hectare with a diameter above 20 cm. It was followed by *Dipterocarpus obtusifolius* with 11 trees per hectare, species of the *Lagerstroemia* genus with 8 trees per hectare, *Sindora cochinchinensis* with 8 trees per hectare, *Dipterocarpus alatus* with 7 trees per hectare, and *Dalbergia cochinchinensis* with 6 trees per hectare.

For a better understanding, compensated values, according to the beta function of the diameter distribution of trees for certain commercial tree species from the Paklay inventory region are given in Table 38 and their graphic demonstration in Fig. 60.

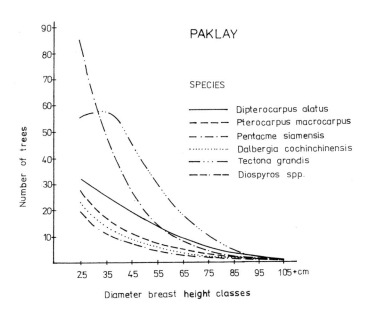

Fig. 60. Compensated values of the diameter distribution of trees for certain tree species according to the beta function. Region of Paklay, Laos.

4.4.4 Possibilities of development of forestry and forest industries

All forests of the Republic of Laos are property of the state. Consequently, the organization and management of forestry are also appurtenant to state organs.

After the reorganization of departments in 1982 the Ministry of Industry, Handicrafts and Forestry came into being. Within the Division of Forestry, sections were established for the education of personnel, for forest inventories, for the protection of the forests and of the living environment, for afforestation, for sawmilling, for other woodworking industries, for state forest enterprises and for the forest service of the provinces. The main concern of the staff of the Ministry is planning, organization and management. Leading personnel in the provinces and state forest enterprises are responsible for other activities, mainly logging, wood processing and afforestation.

The development of forestry and forest industries requires that more information on the country's timber resources are available, that there are more forest inventories and that working plans are prepared. In addition the rational protection of forests must be established in order to prevent soil erosion on degraded land, forest fires and the continuing devastation of forest land by shifting cultivation (Fig. 61). According to FAO data about 100,000 ha of forest is destroyed annually because of deforestation for agricultural crops.

Numerous inventories have been undertaken in Laos. These cover only scattered areas (Persson, 1983). The need for inventory and management information required as a basis for strategy formulation at the national level and for the identification of those specific areas which are available for major industrial investment. There is not yet in Laos forest management. The harvesting of timber on a sustained yield basis from a natural stand requires that the yield should be regulated to prevent over-cutting.

Another task is the planning of logging and a more rational utilization of timber in local consumption, in the local woodworking industry and in foreign trade.

The Laotian Government makes every effort for the development of forestry and forest industries, yet the shortage of export manpower of all categories is a serious obstacle to progress. The Government solves this problem by promoting the education of lower professional staff for forestry and forest industries in the educational institutions established in Laos, and by sending students and trainees to technical schools and colleges abroad.

There are, according to unofficial sources, in Laos about 4 million ha of productive forests with an estimated 70 million m^3 stock of industrial wood. If this estimate is correct, then about 2 million m^3 of logs could be felled in Laos annually. The five-year plan for 1986–1990 envisages an annual log production increasing gradually from 300,000 to 600,000 m^3. At present log consumption in Laos for local demand including house construction is estimated at 200,000 m^3. The annual export from Laos is about 100,000 m^3 of timber and forest products converted to roundwood.

148

Fig. 61. Devastated bamboo forest pushed back by agricultural crops. Region of Luang Prabang, Laos.

Evidently the assumed five-year plan of timber extraction will not be met with, for objective reasons. The actual annual felling of industrial wood does not correspond to the potential.

For the time being, logging is concentrated in regions near the Mekong Valley and near the roads. Logging technology, skidding and trasportation of timber is outdated in most places. The introduction of new technical equipment is required. Modern logging and transportation equipment is used only in some places thanks to foreign technical assistance. The road network is in a poor condition and can be used for log transportation only in the dry season, i.e. 150–200 days in the year.

Compared to other countries in southeastern Asia, the forest industries of Laos are inadequately developed. Woodworking factories are concentrated mainly in the provinces of Vientiane, Savannakhet and Champassak (Fig. 62). The woodworking industry is represented mainly by sawmills. There are at present in the whole country about 55 sawmills with an annual capacity of 450,000 m^3 of logs. Some sawmills are not operational. Other sawmills, though operational, are frequently idle for different reasons: because of the limited and irregular supply of logs, because of the lack of spare parts both for the machine stock of sawmills and for the lorries, and because of the absence of qualified professional staff.

According to Laotian sources the average annual production of sawn wood is 120,000 m^3 (1985), which corresponds to a 50 % utilization of the annual production capacity of the sawmilling industry in Laos. The dimensions of the manufactured

Fig. 62. Timber yard at sawmill. Vientiane, Laos.

sawn wood are governed by usage or are made to order. So far standards for sawn wood production have not been established. About 16 tree species – from among the at least 80 utilizable species – are converted into sawn wood. Most sawn wood is sawn from species of the *Anisoptera* and *Dipterocarpus* genera. Some sawmills use for sawn wood production the so-called luxury tree species such as *Afzelia xylocarpa, Pterocarpus macrocarpus* and *P. pedatus* as well as species of the *Dalbergia* genus.

One veneer and plywood factory has been established in Vientiane. Its annual capacity is 25,000 m^3 of plywood of 4 mm thickness, but at present production is lower. For veneer sheet production mainly *Anisoptera cochinchinensis,* species of the *Dalbergia* and *Pterocarpus* genera, *Chukrasia tabularis, Sindora cochinchinensis* and *Tectona grandis* are used.

Of importance in the woodworking industry are furniture factories, the largest one of them being in Vientiane. The furniture is manufactured for the domestic market and partially also for export, for example small round tables made from luxury timber are being exported to Japan.

Present techniques and technologies used int the woodworking industry are not sufficient for the tasks which the woodworking industry needs to perform in the development of the national economy of Laos. In the interest of further develop-

150

ment of forest industries it is necessary to improve the supply of timber from local resources. This requires the construction of a network of roads, the education of professionals and specialists for the management of forests and forest industries, and the general improvement of the country's infrastructure. The equipment of saw-milling and other woodworking plants should be modernized in order to improve the quality of products, especially that of sawn wood produced for export.

Timber and other forest products have always been an important source of revenue in the foreign trade of Laos. Before World War II, forest exploitation in Indochina was not very systematic. Production was almost entirely in the hands of small operators or small contractors (Sewandono, 1956). In 1979 the revenue from the export of timber and forest products in Laos in general amounted to 20.6 million US $. In 1981 the value of this export dwindled to 8.7 million US $. In 1979 38,400 m^3 of logs, and 70,500 m^3 of sawn wood were exported from Laos, in contrast in 1981 only 9,900 m^3 of logs and 2,200 m^3 of sawn wood were exported. Some plywood and parquetry is also exported (Ahveninen, 1979). The Government's policy is to restrict the export of logs, but no headway is being made to increase the export of sawn wood and other forest products. The "Lao Export–Import Company" is the sole exporter of timber and other forest products.

The main portion of the commercial species of the Lao forests belongs to the Dipterocarpaceae. Export of dipterocarp logs has been banned or restricted in most producing countries including Laos. But for dipterocarp sawn wood are potential market areas: Japan, Singapore and Thailand. Plywood of dipterocarp species is in demand in India, Japan and the Middle East.

In the People's Republic of Laos the task ahead is to draw up a long-term policy for the external trade of timber and other forest products, and to set up short-term commercial plans. At the same time local producers need to be encouraged to improve both their techniques and the quality of manufactured goods.

4.5 Structure of tropical natural forests in other regions of Africa, Asia and Latin America

Although there is an extensive literature concerning closed natural tropical forests, the data published on the structure of tropical forests from different regions are very heterogeneous and therefore difficult to compare. Empirical data are obtained frequently from small areas, e.g. for 0.10 ha; or for one acre, or again from large areas, e.g. a few tens and hundreds of hectares from different sites.

Empirically obtained inventory data are also of unequal value because they are not determined with the same accuracy. Some data were obtained by indirect measurement, other data by direct measurement over a whole area, or on sample plots starting from a certain diameter breast height of tree (e.g. from 10, 20, 40 cm or even from a diameter of 60 cm). In addition, often it is only commercial tree species that are measured and recorded. Only exceptionally are all tree species in the given forest stand recorded and measured (Glerum, 1962).

Inventories of tropical forests are made more or less extensively, when the total area of inventory sample plots amounts to 1–5% of the total area of the inventory region. In such forest inventories inaccuracies easily occur especially if for reasons of financial and time economies, only an estimation of the growing stock is undertaken (Lötsch et al., 1973).

The tree distribution by diameter breast height and the average number of trees (converted to 1 ha) in the moist tropical forests analysed in the previous chapters, showed a great variability with regard to various regions or various tree species. In general the results obtained in natural tropical forests show an obvious decreasing tendency of the diameter distribution of trees from the smallest to the largest diameter breast heights. However, different diameter structures and heterogeneities in the frequencies of commercial and other tree species were recorded in the investigated localities.

The different behaviour of most tree species can be explained by their reaction to light at young and mature ages. All tree species which are capable of growing to dominant dimensions need light for their development, but in their youth they usually grow in the shade. Light demanding tree species, as a rule, have a smaller tree representation in the smallest diameter classes than do shade-tolerant tree species.

In order to verify results of his own investigations the author used data from inventory projects organized by the United Nations Food and Agriculture Organization (Rollet, 1974) in some tropical regions of three continents. Although preference should be given to original numerical data as opposed to amalgamated data for classes, rather than to interpolated data or to graphic values, nevertheless in this case it was necessary to use inventory data converted to a certain area unit.

In the Brazilian lowland evergreen tropical forests of the Amazon river basin, in the region between Belem and Manaus, on inventory plots with a total area of 193 ha, 21,619 trees were recorded, i.e. 112 trees per hectare (Heinsdijk, 1957). The trees were callipered starting from a diameter of 25 cm and they were classified into diameter classes with an interval of 10 cm (classes: 30, 40, 50 cm, etc.). The diameter of the largest tree, measured above the buttresses, was 230 cm; a total of 38 trees (i.e. 2 trees for each 10 ha) had a diameter breast height greater than 150 cm. From among commercial tree species *Carapa guianensis, Vouacapoua americana, Vochysia maxima, Buchenavia huberi, Swietenia macrophylla* and species of the *Xylopia* and *Pouteria* genera were most common in this forest.

Empirical inventory data converted to an area of 10 ha together with theoretical values computed according to the beta function are shown in Fig. 63.

Inventory data from moist tropical forests in Cambodia, the Malayan Federative State of Sabah, and India also indicate an analogous declining tendency of the diameter distribution of trees. Empirical numerical data of the diameter distribution of trees and basic parameters of the beta function from the mentioned forests are given in Table 39.

152

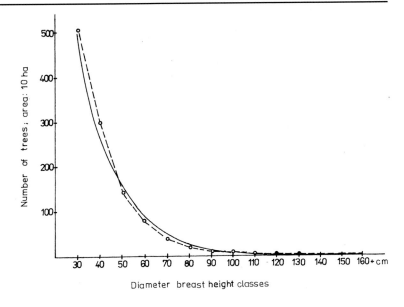

Fig. 63. Empirical and theoretical diameter distributions of trees. Region of Belem, Brazil.

Table 39. Actual data of diameter distribution of trees and their calculated exponents and constant by the beta function. Inventoried regions of Malaysia. Cambodia and India

Midpoint values of D.B.H. classes	Malaysia, Sabah	Cambodia, eastern Mekong	India, western Ghats
cm	Number of trees per ha		
20	96	65	108
30	46	32	56
40	22	15	28
50	15	9	16
60	12	6	10
70	8	3	5
80	4	3	3
90	4	2	3
100+	8	3	4
Total	213	138	233
Const.	0.0091	0.0005	0.0001
α	−0.4409	−0.4254	−0.3660
β	1.5858	2.1298	2.5995
χ^2	16.000	8.059	12.154

In analysing the numerical data from the mentioned regions the tree numbers classified by diameter breast height were converted to an area of 1 ha. In evergreen moist tropical forests in the region of Sandakan in Sabah, commercial tree species of the *Shorea*, *Parashorea*, *Dryobalanops* and *Dipterocarpus* genera were represented most frequently (Burgess, 1961).

Inventory data from Cambodia were derived from the Pralay Triek Forest Reserve in the eastern Mekong region. In this area the commercial tree species *Hopea odorata*, *Shorea vulgaris*, *S. thorelii*, *Vatica dyeri*, *Dipterocarpus dyeri*, *D. turbinatus*, *D. alatus*, *Anisoptera cochinchinensis* and species of the *Lagerstroemia* genus occurred frequently.

From India, inventory data were analysed from the region of evergreen moist tropical forests situated in the Western Ghats Mountains, to the south of Bombay. In these forests *Hardwickia pinnata*, *Myristica athenuata*, *Vateria indica*, *Hopea parviflora*, *Calophyllum tomentosum* and *Dipterocarpus indicus* were the most common commercial tree species. Trees growing to a diameter of 100 cm yield from 3 to 8 m^3 of industrial wood (Menon, 1945).

From among the three mentioned regions the forests in Cambodia are the poorest with regard to the number of trees in the various diameter classes (138 per hectare). A larger number of trees occurs in the evergreen tropical forests of Sabah (213 per hectare) and in India (233 per hectare). Whereas in India, in the Western Ghats Mountains, the forests are characterized by a richer representation of trees in the smaller diameter classes, in the forests of Sabah relatively more trees occurred in the larger diameter classes.

The empirical material from inventories from evergreen tropical forests in the various mentioned localities confirmed previous results concerning the heterogeneity of the composition of natural tropical forests from the point of view of occurrence and diameter distribution of commercial tree speccies. At the same time the clearly declining tendency of the diameter distribution of trees in natural moist tropical forests from the smallest to the largest diameters was confirmed.

While data on the diameter distribution of trees in natural tropical forests can be obtained fairly easily by measuring the diameter or girth at breast height, the measurement of tree heights in natural tropical forests usually is indirect, laborious and rarely performed. For practical reasons in natural tropical forests only the length of the stem from the buttress to the first tree branches is measured. Only seldomly, for informative reasons, are heights of the largest trees of commercial or other species measured.

Because tree distribution by height can provide interesting information for the study of tropical forests, the author used for an analysis the empirical material of the French forester B. Rollet (1974) who measured the heights of all trees in sample plots with an area of 0.25 ha. The measured tree heights were grouped into height classes with an interval $i = 2$ m. Tree heights were measured beginning from a height of 3 m in three localities of evergreen moist tropical forests, namely in Venezuela

Table 40. Empirical data of height distribution of trees in high tropical wet forests of El Dorado (Venezuela), Banco (Ivory Coast) and Mindanao (Philippines). Computed constants and parameters by beta function

Height class	El Dorado	Banco	Mindanao
m	Number of trees; area: 0.25 ha		
4	342	244	264
6	187	122	144
8	122	54	70
10	39	32	46
12	28	22	23
14	23	11	14
16	16	16	17
18	26	6	12
20	12	6	2
22	12	4	7
24	6	2	2
26	10	2	5
28	4	2	6
30	7	1	2
32	4	2	1
34	6	2	1
36	4	2	3
38	3	2	3
40	–	5	1
42	1	4	2
44	1	3	–
46	–	1	1
48	–	–	1
50	–	–	1
52	–	–	–
54	–	–	1
Total	853	545	629
Const.	0.0025	0.0575	0.0000
α	−0.5665	−0.6972	−0.6164
β	2.8254	1.9071	3.7179

in the El Dorado region, in the Ivory Coast in the Banco region, and in the Philippines on the island of Mindanao in the Lianga region.

Tree distribution by height in the natural tropical forest can be expressed mathematically as the diameter distribution of trees. Empirical numerical data from the mentioned three localities in three continents showed that tree distribution by height in the tropical high forest has a declining tendency from the smallest to the largest heights of trees. This regularity is expressed mathematically also by the beta

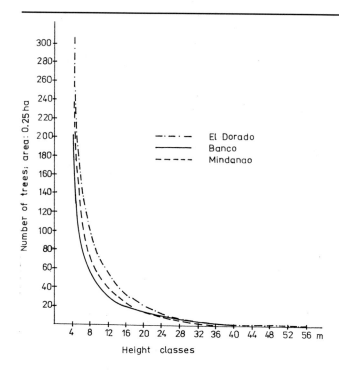

Fig. 64. Height distribution of trees in some regions of evergreen moist tropical forests. The empirical values are compensated according to the beta function.

function. The empirical values of the height distribution of trees in natural tropical forests of the El Dorado, Banco and Mindanao regions and the computed parameters of the beta function are given in Table 40; their graphic illustration in the form of compensated curves is shown in Fig. 64.

For testing the differences between the theoretical and empirical height distributions of trees, the test of good agreement was determined using the criteria according to Kolmogorov and Smirnov (Plokhinskii, 1937). According to the final values of the testing, there are no significant differences between the theoretical and empirical distribution.

4.6 Diameter structure of trees in forests of the miombo type in Tanzania

4.6.1 Basic data

Tanzania, one of the republics of eastern Africa, is situated to the south of the equator between 1°00′ and 11°45′ S and between 29°30′ and 40°30′ E. It borders in the north on Kenya and Uganda, in the west on Rwanda, Burundi and Zaire, in the south on Zambia, Malawi and Mozambique. In the east Tanzania borders on the Indian Ocean.

156

Fig. 65. Sisal plantation with a group of *Chlorophora excelsa* trees. Region of Tanga, Tanzania.

Tanzania consists of the territory of Tanganyika and the islands of Mafia, Zanzibar and Pemba; its surface area is 942,770 km^2 including lakes (51,573 km^2). The population (21 million in 1985) is composed of about 120 different tribes belonging mostly to the Bantu group.

The name of Tanzania came into being on 27 April 1964 when two independent countries – Tanganyika and Zanzibar – established the United Republic of Tanzania with Dar es Salaam as the capital. The official languages are Swahili and English.

Tanzania is an agricultural country. Agricultural land forms 20% of the country's surface. A large part of the country, mainly the central region, is dry and infested by the tse-tse fly and therefore sparsely populated. On the islands of Zanzibar, Pemba and Mafia more than 50% of the surface is agricultural land.

Among agricultural crops sisal (with over 40% of world production) (Fig. 65) and cloves which are cultivated on Zanzibar and Pemba (80% of world production) are of foremost economic importance. Tanzania also exports coffee, cotton, tea, cashew nuts, peanuts, vegetable, legumes, pyrethrum, tobacco, etc. In addition hide, fell and leather, and among minerals diamonds, gold, tin and lead are exported.

Most of Tanzania's territory is formed by a plateau ascending to altitude of 1,200–1,500 m (Figs 66, 67). The central region is very dry and hot. The 20–70 km wide coastal belt is formed by a plain with a moist tropical climate and an average annual temperature of 25–27°C. On Kilimanjaro and on Mount Meru the temperature drops during the night below freezing point. In the country's southern and south-

Fig. 66. Landscape in southwestern Tanzania. Region of Sumbawanga, Tanzania.

Fig. 67. Farmstead in the region of Nyassa, southern Tanzania.

western regions there is one rainy season lasting from December to May. On the remaining territory there are two rainy seasons. The so-called long rainy season lasts from March to June and the period of short rainfall occurs in November and December. On the greater part of the territory the annual rainfall amounts to 750–1,200 mm; on the western shores of Lake Victoria the average annual rainfall reaches 2,060 mm and in the southwestern part of the country in the Tukuyu region it even reaches 2,520 mm. On the island of Zanzibar the average annual precipitation is 1,550 mm and on the island of Pemba it is 1,850 mm.

158

Fig. 68. Afforested land in the region of the northern Kilimanjaro, Tanzania.

Tanzania is a country of natural contrasts. The beaches of the Indian Ocean are in parts interwoven with coral reefs, and around river estuaries mangrove forests occur.

The best-known natural feature in Tanzania is the highest mountain not only of Tanzania, but of Africa – Kilimanjaro – which is situated in the north and close to the border with Kenya (Fig. 68). Kilimanjaro is formed by three extinct volcanos. The highest and youngest one of them is Kibo (5,895 m above sea level) with a rounded and permanently snow-covered summit. To the east of Kibo is the rocky Mawenzi (5,340 m) and the oldest volcano of this mountain, Shiro (4,000 m) is situated to the west of Kibo.

Mount Meru, situated about 60 km to the west of Kilimanjaro, is the second highest mountain in Tanzania having an elevation of 4,568 m and a thousand metre deep crater. The forest-line in these mountains goes up to 3,200–3,500 m above sea level.

To the east of Kilimanjaro stretches the Pare mountain range and farther on there are the Usambara Mountains (Figs 69, 70).

To the west and southwest of Kilimanjaro and Mount Meru lies the dry Masai steppe with smaller cones of extinct volcanos. Farther to the west stretches the mountain range of the Great Craters (Fig. 71).

At the end of the Mesozoic period and at the beginning of the Tertiary period a belt of faults originated in Africa forming the conspicuous natural phenomenon

159

Fig. 69. Devastated forest regions. West Usambara Mountains, Tanzania.

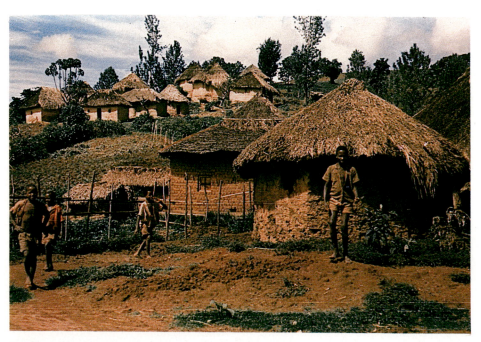

Fig. 70. Settlement in the West Usambara Mountains, Tanzania.

160

Fig. 71. The mountain range of the Great Craters, Tanzania.

of the "Great Rift Valley" which stretches from the Zambezi river in Mozambique to the north along the Lake Malawi. In southern Tanzania the Great Rift Valley divides into two branches. The western branch continues through the lakes of Rukwa, Tanganyika and Kiwu to the Lake Albert. The eastern branch of the Great Rift Valley stretches from the town of Mbeya to Lake Eyasi in the north. From there it continues through Lakes Manyara, Natron and Naivasha to Lake Turkana (Rudolf) and farther north through the Abyssinian lakes and the Red Sea to the Dead Sea (Figs 72, 73).

The mountain range of the Great Craters is noteworthy for the occurrence of several extinct volcanos. The highest peak in this mountain range is Loolmalasin which lies 3,650 m above sea level. The only active volcano in this mountain range, and in Tanzania likewise, is Ol Doinyo Lengai which reaches 2,880 m above sea level. The 2,700 m high and 600 m deep Ngorongoro Crater is the best-known crater in this mountain range. Ngorongoro is one of the world's largest craters; it has an area of 264 km^2. Living in the crater bottom there are about 22,000 wild animals. A vast part of the Great Crater Mountains, including Ngorongoro Crater, has been declared to be a conservation area with a special statute. In this region, together with the strictly protected wildlife, there live about 10,000 Masai with their herds which number about 100,000 head of cattle. In the extensive savanna of the Serengeti there are about 120 different species of game animals, including 35 species of large steppe animals and over 400 species of birds. The Serengeti is the oldest national park in

Fig. 72. The Great Rift Valley in the region of Mbeya, Tanzania.

Fig. 73. Landscape in Tanzania. Mount Mbeya (2,820 m above sea level) in the background; decorative shrubs of *Euphorbia pulchera* in the foreground.

Tanzania (1951), it extends to the west from the Great Crater Mountains and to the east of Lake Victoria (Figs 74–79).

Extinct volcanos also occur in other parts of Tanzania. Thus Mount Rungwe with an elevation of 2,960 m is situated in the south near Lake Malawi, and in the northern part of the central region Mount Hanang towers to an altitude of 3,420 m (Fig. 80).

There are several great lakes in Tanzania, including Lake Tanganyika, the world's second deepest lake. Situated at an altitude of 780 m it is 1,435 m deep and has a surface area of 31,900 km².

162

Fig. 74. Zebras in the Serengeti National Park, Tanzania.

Fig. 75. Feeding giraffe. Manyara National Park, Tanzania.

Fig. 76. Leopard in the thicket of Serengeti National Park, Tanzania.

Fig. 77. Cape buffaloes in the Mikumi National Park, Tanzania.

Fig. 78. Lioness on the morning prowl. Serengeti National Park, Tanzania.

Fig. 79. Lion on the morning prowl. Serengeti National Park, Tanzania.

Fig. 80. Dry deciduous tropical forest. The extinct Hanang volcano, 3,420 m above sea level, in the background. Tanzania.

Fig. 81. Boulders on the banks of Lake Victoria. Region of Mwanza, Tanzania.

Fig. 82. Dry thorny tropical forest with *Acacia pseudofistula* as the main tree species. Serengeti National Park, Tanzania.

Fig. 83. Cut-over area prepared for the establishment of forest plantations with standards of *Chlorophora excelsa*. Rondo Plateau, southern Tanzania.

On the borders of Uganda, Kenya and Tanzania Lake Victoria with a surface area of 68,800 km^2, the largest lake in Africa, is situated (Fig. 81). It is situated between the two branches of the "Great Rift" at an altitude of 1,130 m. The lake reaches a depth of 80 m. All the lakes are economically important for fisheries and as navigable waterways.

Tanzania's rivers are short and mostly flow into the Indian Ocean. The longest river is the Rufiji with its affluent the Great Ruaha. Obviously the most interesting river in Tanzania is the Kagera river which empties into Lake Victoria thus feeding the Upper Nile.

A large part of Tanzania is covered by tree savanna and dry tropical woodland savanna with an area of 117,090 km^2. The term woodland signifies an open forest where more than 50% of the soil is shaded by tree crowns. The largest group of dry tropical deciduous forests is formed by the so-called "miombo" forest.

On vast extents bush and savanna occur with shrub vegetation and scrub which include thorny trees such as *Acacia mellifera* (VAHL.) BENTH., *A. senegal* (L.) WILLD., *A. kirkii* OLIV. and other *Acacia* species (Fig. 82).

The area of closed productive forests in Tanzania attains 130,700 km^2 which

166

equals 14% of the country's territory. Economically the most important are the hight natural forests which have an area of 9,310 km². They mainly represent mountain forests, the most valuable of them being situated in the region of Mount Meru, Kilimanjaro and the Usambara Mountains. These forests occur in altitudes of above 1,200 m where there is an average annual rainfall of 900–2,000 mm. The coniferous species *Juniperus procera* HOCHST. ex ENDL., *Podocarpus usambarensis* PILGER, *P. milanjianus* RENDLE, and broad-leaved species of *Ocotea usambarensis* ENGL., *Cephalosphaera usambarensis* WARB. and *Olea welwitschii* (KNOOL) GILG. et SCHELLENB. are of the greatest economic importance.

In river valleys and at the foot of mountains there are lowland moist tropical forests. A few of them remain because in the past they have been excessively exploited, destroyed by fire in shifting cultivation and partially also gradually transformed into sisal plantations. The most valuable tree species in the lowland tropical forests of Tanzania are *Chlorophora excelsa* (WELW.) BENTH. et HOOK. f., *Khaya nyasica* STAPF ex BAKER f. and *Bombax rhodognaphalon* K. SCHUM. (Fig. 83).

Since timber requirements are increasing in Tanzania, and most domestic tree species grow slowly, foresters started to establish plantations of exotic, mainly coniferous species, such as *Pinus patula* SCHL. et CHAM., *P. radiata* D. DON., *P. caribaea* MORELET. and *Cupressus lusitanica* MILL. Hardwoods have also been planted, mostly *Tectona grandis* (teak) and eucalypts, especially *Eucalyptus saligna* SM., *E. maidenii* F. V. MUELL. and *E. globulus* LABILL.

4.6.2 Growth possibilities of trees in the miombo type of forests

The miombo type of forests are dry open deciduous tropical forests on hills and plateaus which in Africa grow to the south of the equator. The term miombo is derived from a local name for the tree *Brachystegia wangermeeana* DE WILD. which is one of the typical tree species for such forests in the Tabora region in central Tanzania.

The miombo forests cover extensive areas of Tanzania, Mozambique, Zambia, Zimbabwe, Botswana, Malawi, Zaire and Angola. They occur on various soils, including lateritic soil, at altitudes ranging from 700 to 1,800 m and where the annual precipitation ranges from 700 to 1,500 mm. The trees shed their foliage at the beginning of the dry season which lasts from 5 to 7 months. They grow to a height of 7–25 m and to a diameter of 30–60 cm. The stems are short and do not exceed 8–10 m.

The miombo forests have 2–3 storeys. In the main storey tree species of the *Brachystegia, Isoberlinia, Pterocarpus, Afrormosia, Afzelia, Burkea, Xeroderris, Erythrophleum* and *Sterculia* genera occur. Below them is a storey of smaller trees belonging to the *Uapaca, Strychnos, Monotes, Diospyros, Faurea, Pseudolachnostylis, Terminalia, Ostryoderris* and *Combretum* genera. The lowest storey is composed of various shrubs, herbs and grasses.

The miombo forests include a number of tree communities. In Zambia and Zimbabwe *Baikiea plurijuga* is the main economic tree species. In other countries tree species of the *Brachystegia* and *Isoberlinia* (Syn.: *Julbernardia*) genera form the most important communities of the miombo forest. Thus around the border region between Tanzania and Zambia *Brachystegia floribunda* and *Isoberlinia paniculata* prevail, in the central region of Tanzania *Brachystegia spiciformis* and *Isoberlinia globiflora* predominate, in the region of Kigoma near Lake Tanganyika *Brachystegia longifolia* and *Isoberlinia angolensis* are prominant and in Zaire in the region of Lubumbashi *Isoberlinia paniculata* and *Brachystegia spiciformis* var. *latifoliolata* (Malaisse, 1978) are the principle species.

The most valuable tree species of the miombo forests growing to saw log dimensions include *Pterocarpus angolensis*, *Afzelia quanzensis*, *Afrormosia angolensis*, *Albizia versicolor*, *Burkea africana*, *Baikiea plurijuga* and species of the *Brachystegia* and *Isoberlinia* genera (Fig. 84).

Fig. 84. *Brachystegia spiciformis* as a shading tree species. Region of Tabora, Tanzania.

The miombo forests suffer under shifting cultivation, by which, each year regularly part of the forest is destroyed and the growing timber stock is burnt. The forest land thus cleared is temporarily cultivated for agricultural crops (subsistence agriculture). After relinquishing the land because of low yield the natural succession of vegetation starts. Fires kindled almost regularly at the start and at the end of the dry season causes severe damage to the ecosystem. The forest vegetation is destroyed, the soil is impoverished and its fertility is impaired by the destruction of humus and the microflora. Certain tree species with thick bark are resistant to forest fires, e.g. *Erythrophleum africanum, Parinari curatellifolia, Pterocarpus angolensis* at an older age, *Strychnos spinosa, Uapaca nitida* and species of the *Combretum* genus. Medium resistant tree species include species of the *Brachystegia* and *Isoberlinia* genera, *Strychnos pungens, Uapaca pilosa, U. kirkiana, Xylopia odoratissima, Bridelia cathartica* and others.

The number of trees per unit area varies greatly according to the different localities. According to Malaisse (1978) there are in the miombo forests in the Shaba region in Zaire about 65 trees per hectare in the upper (14–22 m high) storey, about 80 trees per hectare in the intermediate (8–14 m high) storey, and 385–500 trees per hectare in the lower storey reaching a height of 8 m.

In the permanent sample plot No. 25 in the Simbo Forest Reserve in the Tabora region in Tanzania, 104 different tree species of various height and age were identified by a detailed inventory in 1970. The sample plot had an area of 3.30 ha and was situated at an altitude of 1,200 m. The average annual precipitation attains 900 mm and is concentrated in the rainy season (December to May). The temperature is around 13–18°C in the cool season (July to August) and over 32°C in the hot season (October to November). In the entire area 21,597 trees and shrubs were counted, including seedlings, this equals 6,545 trees and shrubs per hectare. Of all trees and shrubs 81% were of a height of 0.1–2.5 m. The remainder – 4,105 trees on the plot, i.e., 1,244 trees per hectare (19% of all trees) were of a height of 2.5–25 m (Borota, 1971a).

In the lower stand storey species of the *Combretum* genus prevail, representing 17% of all trees and shrubs. In the intermediate storey *Terminalia sericea* and *T. kaiseriana* were most frequent with a proportion of 3% of all trees in the area. Among commercial tree species *Burkea africana* (6% of all trees), *Brachystegia spiciformis* (4%), *Isoberlinia globiflora* (2%), *Afzelia quanzensis* (0.8%) and *Pterocarpus angolensis* (0.6%) were represented most.

The greatest tree height was attained by *Brachystegia spiciformis* and *Burkea africana* (both 25 m), *Pterocarpus angolensis* (22 m), *Xeroderris stuhlmannii* (20 m) and *Isoberlinia globiflora* (19 m). The largest diameter breast height was measured in *Brachystegia spiciformis* (64 cm), *Burkea africana* (62 cm), *Isoberlinia globiflora* (54 cm), *Xeroderris stuhlmannii* (44 cm) and *Pterocarpus angolensis* (43 cm).

On another permanent sample plot (No. 26) in the Simbo Forest Reserve with an area of 4.20 ha all trees of the main commercial species in the miombo forest were marked so that their diameter increment could be studied.

The sample plot is situated at an elevation of 1,220 m above sea level. The soil is grey-brown and sandy, and the thin humus layer is consumed to a considerable degree by white ants. This miombo forest is represented by the *Isoberlinia–Brachystegia* community. In the upper storey of the stand *Brachystegia spiciformis, B. longifolia, Isoberlinia globiflora, Afzelia quanzensis, Pterocarpus angolensis, P. chrysothrix, Burkea africana, Albizia antunesiana* and *Sterculia quinqueloba* were represented; in the intermediate and lower storeys of the stand *Lannea schimperi,*

Brachystegia wangermeeana and species of the *Combretum, Strychnos, Vitex, Hexalobus, Schrebera, Strophanthus* and other genera occurred. Young trees and seedlings were almost destroyed or damaged by forest fires.

During his prolonged stay in Tanzania the author had the opportunity to take measurements of the diameter breast height of trees of species growing on this plot. There was a total of 552 trees in the plot, or 131 trees per hectare with a diameter of 10–60 cm. The diameter breast height of an ill-shaped over mature stem of *Brachystegia spiciformis* was 82 cm. The tree species *Isoberlinia globiflora, Brachystegia spiciformis* and *Pterocarpus angolensis* were evaluated separately. The remaining tree species were merged into the group "other tree species".

The empirical values of the diameter distribution of trees for the main species and for the whole stand population on plot No. 26, together with their main statistical characteristics are given in Table 41.

Table 41. Empirical data of diameter breast height distribution of trees. Type of forest "Miombo". Experimental plot No. 26, Simbo

Diameter breast height	Empirical data of tree numbers				
cm	*Isoberlinia globiflora*	*Brachystegia spiciformis*	*Pterocarpus angolensis*	Other species	Total
10	33	23	28	31	115
15	23	17	18	17	75
20	31	28	14	15	88
25	40	30	10	14	94
30	39	11	4	9	63
35	23	4	7	7	39
40	26	6	4	11	49
45	9	1	2	3	15
50	7	–	1	–	8
55	1	1	1	–	3
60	1	1	–	–	2
80₊	–	1	–	–	1
Total	233	123	89	107	552

The frequency distribution of all trees by their diameter breast height on the sample plot in the miombo forest shows a declining tendency from the smallest to the largest diameters. The relatively lower representation of trees in the smaller diameter classes can be ascribed to the negative influence of forest fires. Young trees with a thinner bark are less resistant to fire than older trees with a thicker bark.

In the tropics, forests with poor nutrient resources frequently occur. As an example of this, basic data on the growth of certain tree species in the miombo forest in the Geita region in Tanzania are given. Average annual precipitation in this region amounts to 800 mm, the maximum temperature exceeds 40° C and the dry season lasts 8 months. The experimental plot (No. 269) with an area of 0.10 ha is situated at an altitude of 1,300 m.

This is a rather dry type of the miombo forest on poor sandy soil where a limited number of tree species occur with very small diameter increments. All trees on this permanent increment plot were measured in 1967 and 1970. The trees were measured upwards from a diameter breast height of 2.5 cm. The total number of 134 trees on the experimental plot for increment observation included 53 trees (40%) of *Brachystegia boehmii* and 35 trees (26%) of *Brachystegia spiciformis*. Trees of these species attained sporadically 20 m and more of height. Other tree species were rare in the plot; in all, 16 tree species were recorded. Its stand is not advanced as regards diameter growth because maximum diameters breast height only reach values of 35–36 cm *(Brachystegia boehmii* and *B. spiciformis)*. Only 3 trees of *Isoberlinia globiflora,* 1 of *Pterocarpus angolensis* and 3 species of the *Combretum* genus were recorded on the plot.

The current annual diameter increment of the various tree species, and also that of the whole stand population, computed for a three-year period, has very low values, especially for species of the *Diospyros* genus ($\Delta \bar{d} = 0.2$ mm) but also for the more frequent species of the *Brachystegia* genus ($\Delta \bar{d} = 0.7$ mm).

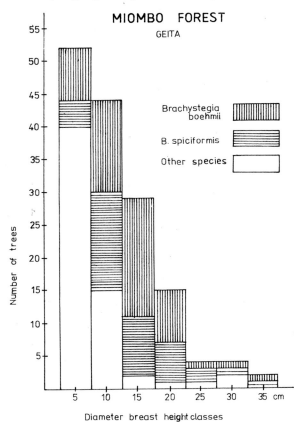

Fig. 85. Diameter distribution of trees in a dry deciduous tropical forest of the miombo type. Region of Geita, Tanzania.

171

The tree distribution by diameter breast height on the experimental plot is shown in the form of a histogramme in Fig. 85. The stand population, considered as a whole, has a clearly declining tendency from the smallest to the largest diameter breast height. In the class of the smallest diameter (D.B.H. = 5 cm) relatively few trees of the *Brachystegia boehmii* and *B. spiciformis* species were recorded. This is because of the destruction of less resistant seedlings by forest fires. The tree distribution of *Brachystegia boehmii* by diameter breast height corresponds approximately to the normal frequency distribution according to Laplace–Gauss. This is confirmed also by testing the homogeneity of the population using the χ^2 test of good agreement. The resulting value $\chi^2 = 9.9$ to six degrees of freedom did not exceed the critical value of $\chi^2 = 12.6$. The tree distribution by diameter breast height of the whole stand population on the experimental plot No. 269 can be expressed by the well-proved mathematical formula, the beta function. This refers to the clearly declining tendency with regard to increasing tree diameter. The resulting value of the test $\chi^2 = 3.9$ to six degrees of freedom did not exceed the critical value.

Interesting figures were obtained by the author from one of the oldest experimental plots for the observation of increment in the miombo forests of Tanzania. This was experimental plot No. 299 in the Ichemba Forest Reserve (Fig. 86). This experimental plot situated at an altitude of about 1,100 m was established in 1931. On the permanent experimental plot all trees of the two most frequent valuable tree species

Fig. 86. Miombo forest with *Pterocarpus angolensis* and *Afzelia quanzensis* as the main tree species. Experimental plot No. 299, Ichemba Forest Reserve, Tanzania.

– *Pterocarpus angolensis* and *Afzelia quanzensis* – were marked. The intention was to measure the diameter breast height of the tress of these two species in five-year intervals.

But, later on, this method of measurement was not adhered to. The author obtained original figures from the measurement of diameter breast height of the mentioned tree species taken in 1931 and after 36 years (1967), he identified the originally numbered and marked trees and again took measurements of their diameters breast height. From the original 85 trees of *Pterocarpus angolensis* after 36 years, 64 trees remained on the plot, and from the original 64 trees of *Afzelia quanzensis* 44 trees were identified in 1967.

The empirical numerical data from the measurement of tree diameters in 1967 were compared with data from the 1931 measurement. The computed characteristics for both tree species are given below. The graphic representation of the frequency distribution of trees by diameter breast height from both measurements for *Afzelia quanzensis* is shown in Fig. 87.

Pterocarpus angolensis, Ichemba Forest Reserve

Measurements taken in	1931	1967
Number of trees on the plot	64	64
Mean diameter breast height $(\bar{d})$ – cm	25.6	34.0
D.B.H.$_{min.}$ – cm	15.2	19.1
D.B.H.$_{max}$ – cm	57.7	65.8
Standard deviation s_d – cm	8.15	9.44
Variation coefficient v_c – %	31.78	27.76

Current annual diameter increment:

$\Delta\bar{d}$ – mm   2.3,

$\Delta\bar{d}_{min.}$ – mm   0.2   (a tree with D.B.H. = 26.6 cm),

$\Delta\bar{d}_{max.}$ – mm   5.4   (a tree with D.B.H. = 51.1 cm).

Afzelia quanzensis, Ichemba Forest Reserve

Measurements taken in	1931	1967
Number of trees on the plot	44	44
Mean diameter breast height $(\bar{d})$ – cm	28.4	36.1
D.B.H.$_{min.}$ – cm	16.3	19.8
D.B.H.$_{max.}$ – cm	43.2	59.7
Standard deviation s_d – cm	8.09	10.28
Variation coefficient v_c – %	28.4	28.5

Current annual diameter increment:

$\Delta\bar{d}$ – mm $\qquad$ 2.1,

$\Delta\bar{d}_{min.}$ – mm $\qquad$ 0.1 $\qquad$ (a tree with D.B.H. = 21.6 cm),

$\Delta\bar{d}_{max.}$ – mm $\qquad$ 5.4 $\qquad$ (a tree with D.B.H. = 56.4 cm).

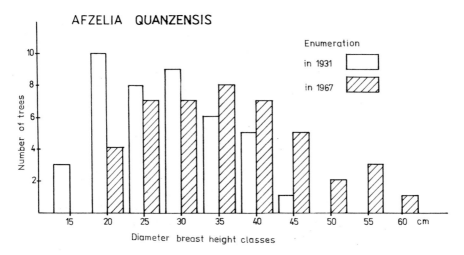

Fig. 87. Diameter distribution of trees for *Afzelia quanzensis* in a miombo forest. 1931 and 1967 inventories. Ichemba Forest Reserve, Tanzania.

The tree distribution for both species on the experimental plot has been considered also as the function of diameter increment (Δd), broken up into classes with an interval of 1 mm (Table 42).

Table 42. Distribution of trees in the "Miombo" type of forest, according to diameter increment. Experiment 299, Ichemba

D.B.H. increment Δd	Number of trees	
mm	*Pterocarpus angolensis*	*Afzelia quanzensis*
1.0	17	22
2.0	24	7
3.0	13	6
4.0	6	2
5.0	4	7
Total	64	44
$\Delta\bar{d}$, mm	2.3	2.1

The distribution of the number of trees with regard to their diameter increment is heterogeneous and uneven. It does not correspond to the theoretical normal distribution, nor to the beta distribution.

174

The mean diameter increment of both tree species ($\Delta \bar{d}$ = 2.1 and 2.3 mm) is very small, when compared to the growth of tree species in evergreen tropical forests, and it is also smaller than the diameter increment of tree species on average site classes in the temperate zone.

One of the richest regions of miombo forests in Tanzania with regard to the occurrence of the very valuable commercial tree species of *Pterocarpus angolensis* (muninga), is situated in the western part of Tanzania in the proximity of the town of Npanda. About 30 km to the southeast from this town, in the Sikitiko Forest Reserve a permanent experimental plot (No. 475) was established for the study of the diameter increment of trees of the species *Pterocarpus angolensis*.

This experimental plot has an area of 2.71 ha, it is situated at an altitude of about 1,050 m. The soil is sandy to loamy and of grey colour; the average annual precipitation amounts to 1,020 mm; the rainy period lasts from December to the end of May.

The author took measurements of the diameter breast height of the trees growing on this plot in 1967, he evaluated these numerical data and compared them with data recorded in measurements taken in 1962. He evaluated the diameter increment of trees for a period of five years. The graphic representation of empirical values of the diameter distribution of trees of the species *Pterocarpus angolensis* is shown in Fig. 88.

On the permanent experimental plot in Sikitiko measuring 2.71 ha, a total of 236 trees of the species muninga with a diameter breast height greater than 7.5 cm (i.e. 87 trees of muninga per hectare) were recorded at the inventory made in 1967.

As is evident from the graphic representation (Fig. 88) the tree populations of muninga correspond as for the character of diameter breast height, approximately to the normal distribution of frequency according to Laplace–Gauss. It should be emphasized that such a distribution of frequency of trees is unusual in a natural unevenly aged forest. It is well-known that the ecosystem of the miombo forest is poor in nutrients. Nevertheless this system is able to maintain its productivity in unimpaired conditions (shifting cultivation) owing to various nutrient conserving mechanisms functioning in the humus layer and in the root layer above the mineral soil.

The unusually small number of trees in the thin diameter classes may be attributed to forest fires. The miombo forests are relatively resistant to fire (especially as regards mature trees with a thick bark, such as muninga has), but young trees and seedlings are not fire resistant. Forest fires cause damage to the ecosystem, they destroy the microflora, they heavily impair stand regeneration and diminish the fertility of the soil (Fig. 89).

The largest muninga trees on the permanent experimental plot in Sikitiko were 20–22 m high. The current annual diameter increment of *Pterocarpus angolensis* computed from the difference between two inventories with a five-year interval on 236 trees, varied across a range of 1–9 mm. The average arithmetical mean of the annual diameter increment ($\Delta \bar{d}$ = 2.2 mm) is relatively low. It corresponds to the

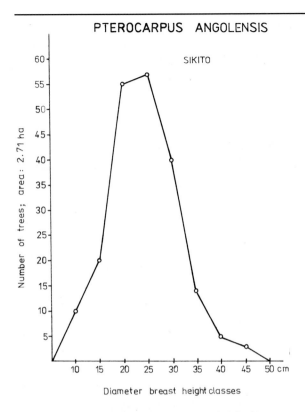

PTEROCARPUS ANGOLENSIS

SIKITO

Number of trees; area: 2.71 ha

Diameter breast height classes

Fig. 88. Diameter distribution of trees for *Pterocarpus angolensis* on the experimental plot No. 475 at Sikitiko Forest Reserve, Tanzania.

mean annual diameter increment of *Afzelia quanzensis* and *Pterocarpus angolensis* in the miombo forest on the permanent experimental plot in Ichemba.

The tree distribution of muninga on the experimental plot No. 475 in Sikitiko was studied by the author, also with regard to the diameter increment (Δd) at breast height. In addition to the verbal description, as an objective approach the empirical and theoretical values of the tree distribution (according to the beta function with regard to the diameter increment as the typical character) are graphically represented in Fig. 90.

Figure 90 confirms that the beta function is suitable for the mathematical expression of the tree distribution of *Pterocarpus angolensis* in the miombo forest with regard to diameter increment. There is a clearly declining tendency in the tree distribution from the smallest to the largest values of the diameter increment.

The smallest annual diameter increment ($\Delta d = 1$ mm) was observed in 110 trees, i.e. 47% of all trees on the plot with a diameter breast height of 10–40 cm. The largest annual diameter increment ($\Delta d = 9$ mm) occurred in the tree with the largest diameter on the plot D.B.H. = 44.7 cm. In 162 trees, i.e. 69% of all trees on the experimental plot, an annual diameter increment of 1–2 mm was observed. These data draw attention to the great diversity in the range of the diameter increment, and consequently also to the volume increment of trees in natural dry deciduous forests of the miombo type. They show, at the same time, how small are the incre-

176

Fig. 89. Sikitiko experimental plot No. 475, Tanzania.

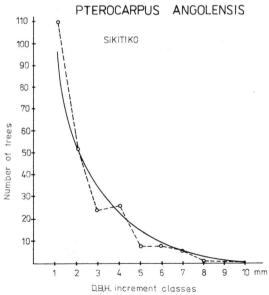

PTEROCARPUS ANGOLENSIS

SIKITIKO

Number of trees (y-axis: 10–110)

D.B.H. increment classes (x-axis: 1–10 mm)

Fig. 90. Declining tendency of the diameter increment in *Pterocarpus angolensis*. Experimental plot No. 475, Sikitiko Forest Reserve, Tanzania.

177

ments of trees and how low the production of timber is in such conditions. They emphasize, too, the great variability of the tree distribution for various species and in whole stand populations with regard to the diameter breast height.

The more important economic tree species in the dry deciduous tropical forests of the miombo type in Tanzania are as follows (Bryce, 1967):

Afrormosia angolensis (BAKER) HARMS. – muvanga

Afzelia quanzensis WELW. – mkora

Albizia versicolor WELW. ex OLIV. – mtanga or musase

Brachylaena hutchinsii HUTCH. – muhuhu

Brachystegia boehmii TAUB. – mkuti

B. spiciformis BENTH. – mtundu or messassa

Burkea africana HOOK. – burkea or mgando or mukarati

Dalbergia melanoxylon GUILL. et PERR. – mpingo or blackwood

Dialiopsis africana RADLK. – mkalya

Diospyros mespiliformis HOCHST. ex A. DC. – mgiriti

Erythrophleum guineense G. DON. – missanda or tali

Julbernardia globiflora (BENTH.) TROUPIN. – muwa

(Syn.: *Isoberlinia globiflora* (BENTH.) HUTCH. ex GREENWAY)

Millettia stuhlmannii TAUB. – panga-panga

Pteleopsis myrtifolia (LAWS) ENGL. et DIELS. – mwindi

Pterocarpus angolensis DC. – muninga

Pygeum africanum HOOK. f. – mueri

Spirostachys africana SOND. – msaraka

5 Structure of artificially established forests in the tropics

The growth process of even-aged forest stands or forest plantations, as artificially established stands in the tropics and subtropics are often called, is different from that found under natural conditions. The tree distribution by diameter breast height as a function of age is distinct in the artificial plantations where the most developed part of the growing stock increases its volume and forms part of the stand until its maturity. A certain number of trees terminate further growth because of the lack of nutrients and space, and they therefore perish. In even-aged stands (plantations) such trees are usually removed by thinning operations together with other yet living trees which, though still producing wood, hinder the development of trees which have better growth increments and which are of a better quality because of their position in the stand.

The numerical dwindling of trees in artificially established even-aged stands is a result of their natural and artificial origin. With increasing age the trees differentiate ever more in diameter, height, volume and increment, and participate in timber production according to their hierarchical position in the forest stand.

It is well-known from investigations by many authors (e.g. Cajanus, 1914; Näslund, 1936; Prodan, 1951 – cited by Borota, 1960), that a forest stand consisting of even-aged individuals of the same species corresponds approximately to the normal distribution of frequencies according to Laplace–Gauss with regard to a certain character, e.g. the diameter breast height (D.B.H.) of trees. Diameter distribution of even-aged forest stands were studied in Europe during the last century. With the increased application of statistical methods also forest investigators observed the close resemblance between diameter distributions of even-aged forest stands and normal curve. This fact gives to the normal distribution its important place in statistics. According to Meyer (1953), the first successful analysis of diameter distributions was made by the Finish forester Cajanus in 1914 and in the United States by Baker in 1923 and Bruce in 1926.

The normal distribution of frequencies can be determined by the arithmetical mean and the standard deviation and expressed mathematically by the function:

$$y = \frac{1}{s_d \sqrt{2\pi}} \, e^{\frac{1}{2}\left(\frac{d - \bar{d}}{s_d}\right)^2},$$

where y is the ordinate of the point of the curve, d the diameter breast height of the tree, $\bar{d}$ the average arithmetical mean, s_d the standard deviation around d, e the base of natural logarithm (e = 2.71828), π (= 3.14159).

This mathematical function, represented graphically, is bell-shaped and symmetrical to the vertical axis passing through the arithmetical mean.

The development of even-aged forest stands is influenced by a number of rather variable factors which impair the homogeneity of the population, thus preventing the total fulfillment of preconditions required by the Laplace–Gauss regularity. Hence it is desirable to express the tree distribution in a stand by diameter breast height, and height or diameter increment, respectively, the author has used the normal distribution of tree frequencies as a mathematical model. It should be noted that in the conditions of the tropics and subtropics the growth process of forest tree plantations is considerably faster than it is even-aged forest stands of the temperate zone.

Using the example of certain important broad-leaved and coniferous tree species, analyses of the diameter and height distribution of trees together with an analysis of their increment and timber production will now be made.

5.1 *Tectona grandis* L. f., Verbenaceae family

Tectona grandis, commercially called teak, teek, or teca, is one of the most cultivated broad-leaved tree species in the tropics. It grows naturally in the Indian subcontinent, in northern and southern Burma, in the western, northwestern and northeastern parts of Thailand, in the tropical deciduous forests of Indochina and on Java between 10 and 25° N latitude and 73–115° E longitude.

The most favourable growth conditions for teak exist in those tropical climates which have an annual precipitation of 1,250–1,800 mm and a more or less uniform temperature with a minimum of 12°C and a maximum of 38°C. It also grows on drier localities with annual precipitation of around 750 mm. It occurs on various soils, in plains and on hills up to an altitude of 950 m. The best growing conditions for teak are found on deep, fertile and permeable soil at altitudes from 200 to 700 m, where it attains diameters of 50–60 cm and heights of 25–30 m in 60 years (Champion, 1936). According to K. R. Venkatramana Dyer, the greatest height of a teak tree was measured in evergreen forest near the Karumoya river in southern Malabar in southern India, namely 58 m, and according to H. Tireman, the largest diameter breast height was likewise found in southern India in the forests of southern Coorg, namely 245 cm. According to A. Wimbush, the teak tree of the largest volume was probably felled at Palacadava in southern Coimbatore also in southern India which yielded five logs totalling 31 m^3 of timber (cited by Troup, 1921).

Teak prospers on gneiss and granite mother rocks, slates and other metamorphosed rocks as well as on tertiary sand and on limestone. Teak is not widespread on lateritic soil, or in maritime tidal regions, nor in evergreen wet tropical forests with a high rainfall or in very dry tropical regions.

Natural teak forests occur in regions having relatively drier or more humid climate and they are managed accordingly. In moist tropical forests teak is cultivated usually in the high type of forest, in drier conditions the coppice forest, or coppice with standards prevail.

In natural moist deciduous tropical forests in India with an annual rainfall of 1,250–1,650 mm teak forms mixed stands with a 20–60% representation of teak. In such conditions teak occurs together with *Adina cordifolia, Anogeissus latifolia, Dalbergia latifolia, Lagerstroemia lanceolata, Pterocarpus marsupium, Salmalia malabarica, Terminalia tomentosa, T. paniculata* and other tree species. Among bamboo *Dendrocalamus strictus* and *Bambusa arundinacea* occur frequently.

High teak forests are often managed by the system of selective logging with a felling cycle of 30–40 years. On certain localities in mixed teak forests concentrated regeneration fellings with natural or complementary artificial regeneration are also used. On drier sites with less fertile soil coppice teak forests with a rotation of 30–40 years are frequently cultivated (Champion and Griffith, 1960). Wood produced thus is used mainly by the local population. In some degraded forests after clear-cutting, teak plantations are then established.

The timber of teak is very durable, hard, strong, resistant to vermin and it can be easily processed. It is used for ship-building, for the construction of buildings and bridges, for railway carriages and sleepers, for luxury furniture decorative veneer and for wood-carving.

The artificial establishment of teak plantations in tropical regions began in the first half of the 19th century in southeastern Asia. Around the year 1900 the cultivation of teak also began in tropical Africa and today there are teak plantations in almost all tropical countries.

Teak is a light demanding tree species. It needs full illumination for good growth and this should be heeded in the establishment of teak plantations. Because direct sowing of teak seed does not always give satisfactory results, at present teak plantations are established by planting. For example, in southern India this is done by using the so-called stump plants. Strong plants obtained from nurseries, in this case, are not planted as they are, but they are first trimmed by cutting off the part above ground about 5 cm above the plant's neck and by shortening the root to 10–12 cm. Thus a stump plant is obtained which is planted into prepared ground at the beginning of the rainy season. Teak plantations are established usually in a spacing of 1.80 × 1.80 m and up to 2.60 × 2.60 m (Fig. 91).

The best picture concerning the growth and increment of trees and forest stands is obtained by analysing numerical data from inventories on permanent experimental plots. On the territory of today's India and Burma permanent experimental plots in teak plantations were systematically established from 1917 onwards; they served as a source of information on the production and increment of wood in the respective stands. Using numerical data from such permanent experimental plots yield tables for teak were set up.

The yield tables for teak plantations in India and Burma (Laurie and Bakshi Sant Ram, 1940) contain data for the main and secondary stands for total production and total increment. In contradistinction to yield tables used in this country (Czechoslovakia), in the Indian yield tables the top height of the stand, and the volume of stem timber (i.e. the volume of standing trees from a diameter of 20 cm under bark and stem smallwood with a diameter between 5 and 20 cm over bark) are given.

The tables of Laurie and Bakshi Sant Ram also contain data on tree distribution in main stand by diameter classes. Because of this, the author has made a computation of the basic statistical characteristics for the determination of the diameter distribution of trees (Table 43). He has represented this diameter distribution of trees graphically for the first site class at the age of 30, 50 and 70 years (Fig. 92).

Table 43. The statistical characteristics of diameter distribution of trees. Yield tables for teak plantations in India and Burma (1940)

Site classes	Age in years	Number of trees per acre	$\bar{d}$	s_d	v_c	A	E
			cm		%		
I	30	76	30.5	5.91	19.4	0.155	0.027
	50	49	43.0	7.55	17.6	0.144	−0.186
	70	32	60.6	11.02	18.2	−0.561	−0.351
III	30	114	20.4	4.57	22.4	0.378	0.125
	50	73	29.0	5.77	19.9	0.164	−0.066
	70	50	42.6	7.63	17.9	−0.105	−0.168
V	30	301	10.5	2.98	28.4	0.246	0.505
	50	192	14.6	3.55	24.3	0.382	−0.151
	70	144	19.7	4.20	21.3	0.237	−0.197

1 acre = approximately 4,050 m^2.

The number of trees in the stand declines with increasing age. At the same age on a better site class there is a smaller number of trees than on poorer site classes. This is a well-known regularity. But the number of trees per unit area is markedly smaller if compared to forest stands in the temperate zone. The mean arithmetical diameter breast height of the stand $(\bar{d})$ rapidly grows with the age of the stand, progressing approximately equally on all site classes. But the absolute values of the arithmetical mean diameter breast height differ considerably on different site classes at the same age. For example, at the age of 30 the arithmetical mean diameter breast height on the third site class attains 67%, and on the fifth site class 34% of the value of the arithmetical mean diameter on the first site class.

The standard deviation (s_d) has a rising tendency with increasing age on all site classes. Stands on better site classes have a larger diameter variability. The variation coefficient (v_c) as the relative measure of variability indicates a greater variability in younger stands compared with older ones. On poorer site classes the diameter variability of the stand is relatively greater than it is in stands on better site classes. Positive values of the coefficient of asymmetry (A) show a left-hand asymmetric diameter distribution of trees on all site classes in stands up to the age of 50 years. Only older stands on better site classes have a right-hand asymmetric distribution. The frequency distribution of trees is very moderate, even moderately asymmetric.

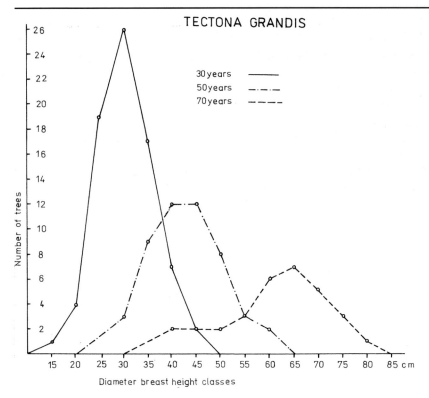

Fig. 92. Diameter distribution of trees in *Tectona grandis* on the first site class according to yield tables set up for India and Burma.

The coefficient of excess *(E)* characterized the pointedness or flatness of the frequency distribution. If the coefficient of excess is negative, the frequency distribution is flat. If the value of the coefficient of excess is positive, the frequency distribution is pointed. According to the yield tables teak plantations at a near-mature and mature age have a flatter frequency distribution of trees than that found in a normal one. Only in younger stands is the diameter distribution of trees usually more pointed than that found in a normal one.

More up-to-date yield tables for teak in India (set up in 1959 under the leadership of S. K. Seth) have four site classes (Seth et al., 1959). The range of the top stand height at the age of 50 years varies between 12 m (fourth site class) and 30 m (first site class).

The rotation of teak plantations for obtaining logs of good quality is usually around 70–80 years. At the age of 80 years, according to the yield tables, the timber volume of the main stand on the first site class amounts to 340 m³ per hectare (= 78 trees) with a total timber production (which includes volume from thinnings) of 637 m³ per hectare. In thinnings 47% of the whole timber stock is exploited. The volume of the mean stem (with a diameter above 5 cm) in a teak plantation on a first

184

site class at the age of 50 years is $2.56\,\mathrm{m}^3$, and at the felling age of 80 years it is $4.36\,\mathrm{m}^3$. With declining site class quality the timber production of stands also declines. On the fourth site class the total production of teak plantations is only $159\,\mathrm{m}^3$ per hectare.

Permanent experimental plots are also established in natural teak forests in India. The economically most important natural teak forests include the Chanda region in the state of Maharashtra in central India. In this region the average annual precipitation amounts to 1,370 mm. Air temperature varies from $3°\,C$ (December to January) to $40°\,C$ (April to May).

In the South Chanda Division in the Allapalli Forest, a permanent experimental plot No. 9 was established in a high teak forest (MacDonald and Mujundar, 1946). Its area is 0.24 ha and it is situated on a plain at an altitude of 160 m. The geological substratum is formed by granite and gneiss. The soil is loamy, dark and partially covered by fallen leaves, herbs and grasses. The most common tree species was *Tectona grandis* with an admixture of *Terminalia tomentosa, Stephegyne parvifolia, Cleistanthus collinus, Dalbergia latifolia* and, in the undergrowth, mainly *Daedalacanthus purpurascens* and *Petalidium barlerioides*.

At the approximate stand age of 71 and 89 years the following data were obtained by inventories:

Age	71 years	89 years
Number of trees per hectare	215	189
Mean stand diameter – cm	27.9	31.3
Top height of the stand – m	22.6	24.5
Basal area – m^2 per hectare	13.2	15.1
Timber stock (D.B.H. $\geq 5\,\mathrm{cm}$) – m^3 per hectare	118	145
Volume of the mean stem – m^3	0.55	0.73

By a thinning at a stand age of 80 years, $12\,\mathrm{m}^3$ of timber per hectare with a diameter of D.B.H. $\geq 5\,\mathrm{cm}$ were extracted. Total current annual volume increment in the observed period amounted to $2.2\,\mathrm{m}^3$ per hectare. This indicates a teak forest of poor productivity.

In the Allapalli Forest Reserve, permanent experimental plot No. 6 was established in a coppice forest. It is situated on a level terrain at an altitude of 180 m, its area is 0.40 ha. Climatic and soil conditions are similar to those mentioned for plot No. 9. The stand originated from coppice sprouts after clear-cutting. The rotation was set at 40 years.

Inventories at the stand age of 23 and 31 years furnished the following growth data:

Age	23 years	31 years
Number of trees per hectare	625	480
Mean stand diameter – cm	14.2	16.3
Top height of the stand – m	15.2	16.8
Basal area – m^2 per hectare	10.1	10.0
Timber stock (D.B.H. $\geq$ 5 cm) – m^3 per hectare	59	64
Volume of the mean stem – m^3	0.09	0.13

A thinning at the stand age of 31 years removed 145 trees per hectare (i.e. 23% of all trees), and 9 m^3 of timber per hectare (i.e. 13% of the timber stock). Total current annual volume increment at the stand age of 31 years was 2.5 m^3 per hectare.

Basic statistical characteristics of the diameter distribution of trees on the experimental plots Nos 6 and 9 in Allapalli are computed for further analysis in Table 44.

Table 44. The statistical characteristics of diameter distribution of trees. *Tectona grandis*. Allapalli, sample plots Nos 6 and 9

Age in years	Number of trees	$\bar{d}$	s_d	v_c	A	E
		cm		%		
Sample plot No. 6						
23	185	14.6	3.11	21.3	0.191	0.051
31	185	16.9	3.44	20.4	0.720	0.810
Sample plot No. 9						
71	48	27.4	5.86	21.4	0.244	−0.669
80	48	29.2	6.32	21.6	0.357	−0.603

The absolute values of the arithmetical mean average and of the standard deviation increase with increasing age. The variation coefficient (which eliminates the influence which the size of the character values has on the size of their dispersion), indicates 1. a slightly smaller diameter differentiation at a higher age of the coppiced teak stand (plot No. 6), and 2. a negligible difference of increase during 9 years in the high teak forest nearing maturity (plot No. 9). The left-hand asymmetric distribution of tree frequencies increased on both experimental plots at a higher age of the stand. The coefficient of excess indicates a larger pointedness of the diameter distribution at a higher age of the coppice stand (plot No. 6), and a diminishing of the moderately flat tree distribution at a higher age of the teak stand which originated from seed (plot No. 9).

Informative data will now be given on the growth of teak in the region of the Nilambur Forest Division in Kerala, in the Aravallikkavu Forest (Fig. 93).

186

Fig. 93. The biggest tree in the world's oldest teak plantation (preservation plot No. 5) aged 117 years at Aravallikkava Forest. Nilambur Forest Division, India.

The protected plot No. 5 is a preserve, it is today a remainder of the oldest teak plantation in the world; the so-called Conolly plantation which was established in 1846. The plot is situated at an altitude of about 50 m. The area of the plot is 2.28 ha. The soil is loamy, deep and fertile, though humus is absent. The average annual precipitation is 2,850 mm and the rainfall is concentrated into the periods of the southwestern monsoon (April–May) and of the northeastern monsoon (August–October). The average daily temperature amounts from 30 to 35°C in March. The relative air humidity is 85–95%.

Some data from this protected stand are given below, according to an inventory made at the age of 115 years:

Site class II	
Number of trees	73 per hectare
Timber stock (stem timber)	497 m³ per hectare
Volume of the mean stem (stem timber)	6.83 m³ per hectare
Height of the tallest tree on the plot	46 m
Length of this stem up to the first branch	24 m
Diameter breast height (D.B.H.) of this stem	127 cm
Volume of the stem up to the first branch	13 m³

In eastern Africa, in Tanzania, teak has been cultivated since 1909. The oldest plantations are in the Kihuhwi area, at an altitude of 200–250 m. The Kihuhwi area is situated about 50 km to the west of the seaside town of Tanga. There were 15 ha of teak plantations in this region in 1916. The climate here is hot and moist, with an annual precipitation of 1,600–1,800 mm.

In one of the oldest teak plantations in Kihuhwi the permanent experimental plot No. 4 (Fig. 94) was established where the following data were obtained by measurement:

Area	0.21 ha
Age of the plantation	61 years
Number of trees	200 per hectare
Mean diameter breast height	40.2 cm
Mean annual diameter increment, Δd	6.6 mm
Mean height of the stand	31.6 m
Mean annual height increment, Δh	0.52 m
Basal area of the stand	25.3 m² per hectare
Timber stock of the stand	312 m³ per hectare
Mean annual volume increment	5.1 m³ per hectare
Volume of the mean stem	1.56 m³

Empirical values of the diameter distribution of trees on the experimental plot at the stand age of 42 and 61 years are represented graphically in Fig. 95.

Fig. 94. *Tectona grandis.* Sixty-year-old plantation. Kihuhwi Forest Reserve, Tanzania.

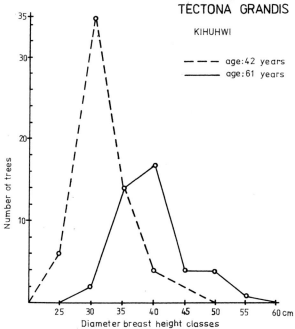

Fig. 95. Diameter distribution of trees on an experimental plot at the age of 42 and 61 years in *Tectona grandis.* Kihuhwi Forest Reserve, Tanzania.

189

5.2 *Shorea robusta* GAERTN. f., Dipterocarpaceae family

Shorea robusta, commercially called sal, is an important commercial tree species in India and Nepal. It forms straight cylindrical stems with a conical on ball-shaped crown which only rarely is entirely without leaves. It usually attains a height of 30–35 m and a diameter breast height of 120 cm. On very good sites it grows to a height of 40–50 m and to a diameter of 200 cm. The tallest recorded sal tree in the region of the Haldwani Forest Division (Uttar Pradesh) was 50 m high, its diameter breast height was 92 cm and the stem volume 16 m^3 (Singh, 1957).

The heartwood is brown, hard, very strong and valuable. The specific weight of the dry wood is 0.80 g per cm^3. The timber of sal is used for the construction of buildings and houses, for the manufacture of railway carriages and sleepers, etc. From the wood resin is extracted which is used as incense in Hindoo temples.

The sal tree occurs naturally in the northern, central and eastern part of India, Buthan and in Nepal. It is distributed roughly in two separate large areas. One area of occurrence of *Shorea robusta* is in the Himalayan foothills from Kangra to Assam (76–93° E longitude). This area borders in the south on the Ganges lowlands. Here the sal tree grows on altitudes up to 1,200 m, sporadically up to 1,500 m. To the south of the Ganges river the second large area of sal distribution is situated, extending throughout central India in the states of Bihar, Orissa, Andhra Pradesh, Madhya Pradesh and Maharashtra (Fig. 96). The species grows on plains and hills up to an altitude of 900 m (Champion, 1936).

Shorea robusta grows on various soils, mostly on loamy-sandy to loamy soil, but also on lateritic soil, e.g. in Bengal. It does not occur on limestone soil. It does best on deep, well-drained loamy to sandy soil.

The average annual temperature in typical sal regions varies from 24 to 27° C. The average minimum temperature of the coolest month – December – is 7 to 13° C, and the average maximum temperature of the warmest month – May – is 34 to 42° C.

The average annual rainfall in the sal area varies from 1,075 mm (Kheri, Uttar Pradesh) to 3,600 mm (Kochugaon, Assam). The average relative air humidity is 60–70%. In Assam sal also occurs in regions with annual precipitation around 5,000 mm.

According to the annual precipitation, sal forests may be classified as follows:
– dry sal forests,
– moist sal forests,
– coastal sal forests, and
– very moist sal forests.

Dry sal forests occur mostly in the Himalayan foothills, in the Siwalik Mountains. These are for the main part mixed forests with a sal representation of 10 to 50% and contain other main tree species such as *Anogeissus latifolia, Acacia catechu,*

Fig. 96. Moist deciduous tropical forest with *Shorea robusta* as the main tree species. Angul Forest Division, India.

Cassia fistula, Cedrela toona, Dalbergia sissoo, Salmalia malabarica, Lagerstroemia parviflora, Terminalia tomentosa, etc. In the lower storey *Colebrookia oppositifolia* and the bamboo *Dendrocalamus strictus* are typical. *Shorea robusta* finds better growing conditions in the valleys where it forms pure stands. At the age of 100 years the trees attain a height of 30–35 m.

Moist sal forests are distributed both in the Himalayan foothills and on the Indian peninsula proper. Annual precipitation does not exceed, as a rule, 1,900 mm. Beside sal which is the main tree species, *Pterocarpus marsupium, Terminalia tomentosa, Eugenia jambolana, Diospyros melanoxylon* and other species are of commercial importance.

The coastal sal forests occur in the state of Orissa. They also include the commercial tree species *Dillenia pentagyna, Terminalia tomentosa, Adina cordifolia, Bursera serrata* and *Xylia xylocarpa.*

In the very moist sal forests in Bengal and Assam *Shorea robusta* grows in communies with *Schima wallichii, Adina cordifolia, Michelia champaca, Gmelina arborea, Sterculia villosa* and other tree species. The undergrowth is formed by ever-

191

green shrubs such as *Colebrookia oppositifolia, Pogostemon plectranthioides,* species of the *Clerodendron* genus and others.

Working circles of sal high forests can be set up according to the silvicultural and management systems in use. They can be approximately of uniform age or they can be of an uneven age with selective logging, and they can also be working circles of coppice or coppice with standards. In any case the regeneration of sal forests is an important task. If natural regeneration of sal is difficult to obtain, reforestation is carried out either by sowing into loosened soil, or, more safely, by planting ball plants cultivated in a nursery. Both sowing and planting is carried out at the begin-ning of the monsoon rains (Ganguly, 1963). It is hard to regenerate sal from seed but the sprouting capacity of sal is very good, therefore in India forests of coppice or coppice with standards are widely cultivated. The wood from coppice forests is used as fuelwood, for charcoal production and for other purposes. The rotation in sal coppice forests is usually around 40 years (Sinha, 1961).

Table 45. Some growth and production data. Yield tables for *Shorea robusta* (1943)

Age in years	Number of trees per ha	Diameter breast height	Top height	Volume of stand	Volume of thinnings	Total volume production
		cm	m	m³ per ha		
Site class I						
50	350	29.2	26.8	234	44	426
70	242	37.3	31.7	309	58	717
100	150	48.3	36.9	366	48	1,104
150	85	64.3	42.4	462	20	1,524
Site class II						
50	410	25.1	22.0	176	32	308
70	285	32.5	25.9	229	43	518
100	172	42.7	30.2	272	34	796
150	97	56.6	34.8	352	14	1,102
Site class III						
50	460	21.1	17.1	115	22	162
70	310	27.7	20.1	149	28	335
100	185	36.3	23.5	177	17	506
150	107	47.5	27.1	240	8	679
Site class IV						
50	507	17.0	12.2	74	14	115
70	327	22.6	14.3	89	19	206
100	195	29.5	16.8	99	10	304
150	117	37.8	19.2	134	4	393

Though coppice forests of sal are widely distributed in India, the cultivation of coppice with standards is more recommended. In such coppice forests trees of the best quality are left as standards and these are held over through two or three rotations of the coppice forest.

In order to obtain empirical numerical data on the growth and increment of sal forest stands, permanent experimental plots were established in India. Results obtained from periodical measurements on these plots served as a basis for the construction of yield tables. Griffith and Bakhshi Sant Ram published in 1943 yield tables for even-aged high sal forests (Griffith and Bakhshi Sant Ram, 1943). The tables include four site classes. The stand site class is the function of age and of the top height of the stand. Some data of the yield tables for sal high forests at the ages of 50, 70, 100 and 150 years are given in Table 45.

The mean diameter breast height of the main stand computed from the basal area of the stand, attains high values in the first and second site classes. If sal trees are to reach a diameter of 40–50 cm then an age of 70–100 years on best sites, and an age of 120–150 years on poorer sites is sufficient. According to the yield tables, total production of stem timber is 1,002 m^3 and of small dimension wood it is 522 m^3, totalling 1,524 m^3 per hectare on the first site class for a rotation of 150 years. The total production of stem timber and small dimension wood is 393 m^3 per hectare for stands on the fourth site class. Thus total timber production in sal high forests varies a great deal and depends mainly on soil conditions. Stands on better site classes are much more developed as to diameter than are stands on poorer sites (Singh, 1957).

The number of trees declines with the age of the stand on all four site classes. The position of the mean diameter breast height increases with age. The actual number of trees per hectare in a main stand on all site classes is smaller when compared to tree species growing in a high forest in central Europe.

More detailed information on the growth of the sal high forest may be obtained from numerical data from the permanent experimental plot No. 6, in the region of the Kheri Forest Division (Table 46).

Table 46. *Shorea robusta*. Growth data in sample plot No. 6. Kheri Forest Division

Age in years	Number of trees	Mean diameter breast height	Basal area	Top height	Volume production
	ha	cm	m^2 per ha	m	m^3 per ha
30	1,275	15.5	23.6	21.3	174
35	947	18.3	24.6	23.8	194
40	565	23.9	24.8	25.3	200
50	447	23.4	28.2	26.3	227
60	367	33.0	31.2	27.7	275
75	262	38.3	29.9	29.9	267

The permanent experimental plot No. 6 was established in the Kheri Forest Division in the Ganganala Forest in the state of Uttar Pradesh. The plot is situated in an approximately even-aged natural high sal forest which forms an irregular quadrangle having an area of 0.08 ha. The site is on a plain, the altitude is 160 m. The average annual precipitation amounts to 1,270 mm which falls mostly in the period of the monsoon rains, i.e. from mid June to late September. The average annual maximum temperature is 30° C, the average annual minimum temperature is 19° C. The soil is loamy, and in the dense undergrowth low trees and shrubs, such as *Mallotus philippinensis, Eugenia jambolana, Ardisia humilis* and various herbs and grass species are represented. The permanent experimental plot was established and measured in a stand which was 30 years old. According to the programme, tree measurements had to be carried out every five years. At the time of establishment the stand was fully closed and was classified, according to top height and age, as belonging to the first site class. There were 63 trees of *Shorea robusta* on the plot which is the equivalent of 1,275 trees per hectare. The sal growing stock was 174 m^3 per hectare and the stand was very dense when compared to the generally accepted

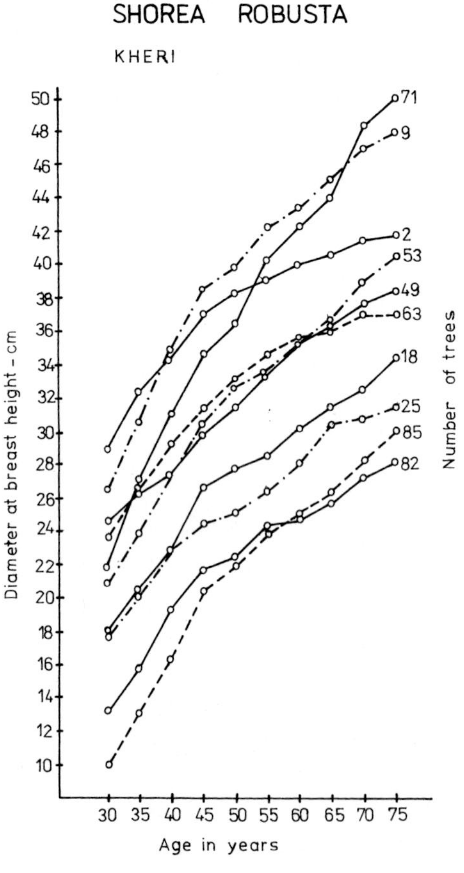

Fig. 97. Development of the diameter breast height as function of the age for ten selected trees of *Shorea robusta*. Experimental plot No. 6, Kheri Forest Reserve, India.

194

yield tables. The number of trees was later decreased by stronger thinning than was originally planned.

The top height of the stand at the age of 30 years was 21.3 m, at the age of 75 years it was 29.9 m. The mean diameter breast height increased during the whole observation period from the age of 30 years (D.B.H. = 15.5 cm) to the age of 75 years (D.B.H. = 38.3 cm). The growing stock on the experimental plot was also recorded from the stand age of 30 years. From total timber production on the experimental plot at the age of 75 years (506 m^3 per hectare) thinnings represented 239 m^3 per hectare, i.e. 47% of the timber production. The growing stock (D.B.H. $\geq$ 5 cm) was 174 m^3 per hectare a the age of 30 years. At the age of 75 years the mean stem volume of the stand was 1.03 m^3.

A shortcoming of permanent experimental plots in tropical forests is their small surface area. On the experimental plot No. 6 in Kheri there were 63 trees aged 30 and 35 years, at the age of 75 years only 20 trees remained. The author selected at random 10 trees from the inventory manual of this plot between the stand ages of 30 and 75 years, and represented graphically the development of their diameter breast heights as a function of age (Fig. 97).

The range of the diameter values of the 10 selected trees was 19 cm (10–29 cm) at the stand age of 30 years, and 22 cm (28–50 cm) at the age of 75 years. The mean annual diameter increment of these 10 trees, considered for the period of 45 years, varied from Δd = 2.9 mm (tree No. 2) to Δd = 6.2 mm (tree No. 71), the average for all 10 trees was $\overline{\Delta d}$ = 3.9 mm. The graphical representation of the diameter growth of randomly selected and measured trees of *Shorea robusta* for a period of 45 years indicates that trees for the main part retain their hierarchical position in the stand during their life-time.

5.3 Tree species of the *Eucalyptus* genus, Myrtaceae family

One of the most interesting groups of fast growing tree species of high yield and use is the *Eucalyptus* genus which belongs to the Myrtaceae family and comprises some six hundred species including natural hybrids.

Most of the eucalypts are confined to tropical and subtropical localities of Australia, but several species also occur naturally on some islands to the north of Australia, and a few species are indigenous to Indonesia and the Philippine Islands (Pryor, 1959).

Eucalypts are light demanding tree species, they are mostly evergreen growing on various types of soil, from poor sandy soil to rich alluvia, on plains and in mountains up to altitudes of 1,500 m. Some species grow in regions of low annual rainfall of 250–350 mm *(E. microtheca, E. salubris)*, others grow where precipitation is up to 4,000 mm *(E. deglupta)*. Some eucalypts reach heights of up to 90 m under favour-

Fig. 98. Fourty metre high trees of *Eucalyptus grandis,* aged 15 years. Arboretum Lushoto, Tanzania.

able ecological conditions *(E. nitens, E. obliqua, E. regnans)* whereas dominant eucalypts in woodlands are widely spaced and reach a height only of 10–20 m *(E. microtheca, E. pauciflora, E. salubris)*. Only a few eucalypts grow in the tropical wet evergreen forests *(E. deglupta, E. urophylla)* (Jacobs, 1979). Eucalypts provide valuable timber and some species are a source of tannin, gum and essential oil.

During the second half of the last century numerous *Eucalyptus* species were introduced into many tropical and subtropical countries. Large-scale plantations of high yielding eucalypts for sawn wood, pulpwood, poles, posts, mine timber, railway sleepers, fuelwood, charcoal and essential oils were established in the twentieth century, especially during the last 30 years. More than 200 *Eucalyptus* species were tested both for timber production and soil conservation. Eucalypts have been planted under a very wide range of conditions including a wide variety of soils and different climates from arid zones with 250 mm annual rainfall *(E. microtheca, E. sideroxylon)* to humid sites with an annual rainfall of 3,000 mm *(E. deglupta, E. urophylla, E. grandis)* and also at various altitudes from near sea level *(E. botryoides, E. grandis)* (Fig. 98) up to elevations of 2,500–3,000 mm *(E. globulus)*.

It has often been found that the fast-growing eucalypts grow even faster in their new homes. A limited number of eucalypts have been used for the establishment of industrial plantations in tropical and subtropical countries. Most of the area of eucalypt plantations established to date as industrial plantations (where the main purpose is to provide raw material for wood processing industries), has been planted with the following species: *Eucalyptus camaldulensis* DEHN., *E. citriodora* HOOK. f., *E. globulus* LABILL., *E. grandis* HILL ex MAID., *E. maideni* F. v. MUELL., *E. paniculata* SM., *E. regnans* F. v. MUELL., *E. saligna* SM., *E. tereticornis* SM., *E. urophylla* S. T. BLAKE and *E. viminalis* LABILL. The total area of eucalypt plantations world-wide is more than four million ha.

Next, information will be given on growth and frequency distribution of trees on permanent sample plots for two economically important eucalypts: *E. globulus* and *E. regnans*.

5.3.1 *Eucalyptus globulus* LABILL.

Eucalyptus globulus, commonly called in Australia "Southern blue gum" or "Tasmanian blue gum", is one of the *Eucalyptus* species of outstanding economic importance which was introduced and widely planted at high elevations in both tropical (Colombia, Ecuador, Ethiopia, India, Peru) and subtropical countries (Chile, New Zealand, Portugal, Spain and Uruguay).

Eucalyptus globulus occurs naturally in Australia, in the relatively cooler southern regions of Victoria and in Tasmania where the annual precipitation is 500–1,500 mm, and it occurs in both coastal and hilly localities at altitudes up to 400 m. The best growth is found on well drained and rich loamy soils where trees reach about 50–60 m in height and 150 cm in diameter.

The timber is light in colour, strong, heavy, moderately durable. The specific weight of the dry wood (moisture: 12%) is 0.60–0.78 g per cm^2. It is used for pulpwood, construction, furniture, poles, particle board, mine timber, fuelwood and for essential oils produced from the leaves. Some information on the growth of this eucalypt outside Australia is given below.

South America

The area of *E. globulus* planted in Ecuador is about 28,000 ha (1980). The best growth occurs in localities between altitudes of 2,000 and 2,900 m with an annual rainfall of 1,000–2,000 mm. Also in Peru *E. globulus* grows fairly well near the equator at an altitude of 3,000 m. The total area of *E. globulus* planted in Chile is about 40,000 ha. The species is planted in coppices with rotations of 10–15 years to provide pulpwood, poles and fuelwood, or it is grown to a felling age of 30–40 years at higher altitudes to provide sawn wood. With 10–15 years rotation the average annual volume increment varies from 10 to 30 m^3 per hectare.

Iberia

E. globulus was introduced to the Iberian Peninsula in the 19th century. It was grown for fuelwood, charcoal, poles and posts.

The total area of *E. globulus* planted in Portugal is about 250,000 ha. Highly suitable sites for planting *E. globulus* exist mainly in a 60–70 km wide belt along the Atlantic coast. In Spain, there are two important regions for planting *E. globulus*: the southwest with an area of about 150,000 ha and the northwest along the Atlantic coast with 120,000 ha. Rotations used for *E. globulus* plantations vary between 8 and 12 years and are almost always based on the coppice system.

According to yield tables for northern Spain the arithmetic mean diameter at 1.3 m varies at the age of 10 years from 10 to 14 cm, and the arithmetic mean height of the stand from 13 to 19 m. The mean annual volume increments obtained vary from 15 to 44 m^3 per hectare. Since the 1960s the wood of *E. globulus* has been chiefly used for paper and textile pulp, particle board, mine timber, parquetry and for essential oil.

East Africa

Nobody knows for certain who first brought the eucalypts to the countries of eastern Africa. It may have been the Arabs or the Portuguese. According to information from British foresters, the planting of eucalypts has been carried out in Kenya during the last half century on land lying between altitudes of 1,500–2,700 m where there is an annual rainfall of 750–1,800 mm. At the beginning of the 20th century *E. saligna* and *E. globulus* were selected as having the best yield. Plantations of these two species were established mainly to provide fuel for locomotives on the railways.

The eucalypt plantations in Tanzania serve to supply poles and fuelwood, to provide shelterbelts and to aid soil conservation (erosion control). The Germans during colonial times planted various species, such as *E. globulus*, *E. botryoides*, *E. microcorys*, *E. saligna* and *E. viminalis*, etc., on some mountain sites in Tanganyika after 1893. At Kigogo Arboretum (which is situated at an altitude of about 1,800 m in Sao Hill with an average annual rainfall of 2,000 mm) the diameter breast height of four measured trees of 33-year-old *E. globulus* varied between 23 and 40 cm, and their heights varied between 32 and 39 m. At Lushoto Arboretum (which lies in the West Usambara Mountains at an altitude of 1,430–1,510 m with an average annual rainfall of 1,070 mm) a plot with 25 trees of *E. globulus* has been measured. At the age of 14 years the mean diameter breast height was 28.5 cm, and the dominant height was 37 m. The soil in the arboretum is a deep red earth.

An experiment conducted in Tanzania on experimental plot No. 250 was established in order to compare a number (i.e. 12) of *Eucalyptus* species with particular reference to the production of transmission poles (Borota and Persson, 1971).

This experimental plot is situated on the western slopes of Mount Meru near Olmotonyi, Arusha region, at an altitude of 1,700 m. The mean annual rainfall for 30 years is 970 mm with a minimum of 640 mm and a maximum of 1,600 mm. The soil is deep volcanic ash, slightly alkaline, and fertile. The previous vegetation was a stand of 25-year-old *Cupressus lusitanica* MILL.

At the age of 13 years all diameters and heights of the 10 largest trees on each plot were measured. The results obtained are given in Table 47.

According to observations of the visible performance of the trees in this experiment, i.e. straightness and branching habit, there are five species which can be classified as suitable for the production of transmission poles. These are: *Eucalyptus maideni*, *E. regnans*, *E. botryoides*, *E. globulus* and *E. saligna* (Fig. 99).

Table 47. Growth data of *Eucalyptus* spp. at the age of 13 years. Olmotonyi

Order	*Eucalyptus* spp.	N	D_G	MDI	DH	DHI
			cm		m	
1	E. maideni	84	25.1	1.92	28.0	2.15
2	E. regnans	51	24.7	1.89	29.8	2.29
3	E. botryoides	86	24.0	1.84	31.3	2.41
4	E. globulus	81	21.1	1.62	31.8	2.45
5	E. saligna	76	21.2	1.63	27.0	2.07
6	E. microcorys	45	23.4	1.80	23.0	1.77
7	E. paniculata	79	21.5	1.65	23.6	1.82
8	E. pauciflora	16	24.3	1.86	15.5	1.19
9	E. citriodora	87	18.6	1.43	24.6	1.89
10	E. resinifera	81	17.8	1.32	23.8	1.83
11	E. maculata	83	19.3	1.48	22.7	1.75
12	E. punctata	111	19.8	1.52	20.2	1.55

Fig. 99. 13-year-old plantation of *Eucalyptus botryoides*. Olmotonyi, Tanzania.

Four further species: *E. microcorys*, *E. paniculata*, *E. citriodora* and *E. maculata* can be classified as having potential for use in pole production, although because of slower growth and crookedness they are not of primary interest.

The other three species are unsuitable for the production of transmission poles. *E. pauciflora* because of very twisted growth and big branches, *E. resinifera* is slow-growing and rather crooked, and *E. punctata* has rather slow growth and poor form.

In this experiment *E. globulus* was distinguished by the fastest dominant height increment, while the diameter growth was considerably smaller than for the three species first listed. Of all species in this experiment *E. maideni* and *E. regnans* have shown the best performance.

In Ethiopia, in the Menagesha region (Addis Ababa), in an 8-year-old plantation of *E. globulus* (established on black clayey soil at an altitude of 2,400 m with an annual rainfall of 1,100 mm), the mean diameter breast height was 17.6 cm, the standard deviation $s_d = 3.5$ cm, and the variation coefficient $v_c = 20.1\%$. The distribution of trees by their diameter breast height corresponded to the normal frequency distribution. The mean height of the stand was 21 m and the growing stock was 214 m^3 per hectare (Berhanu, 1984) using a conversion of 1,310 trees per hectare.

India

In the mountainous conditions of the Nilgiris area in southern India, *Eucalyptus globulus* has a very good increment ability at the age of one hundred years. It may be useful to present some information on the dynamics of growth of this oldest eucalypt plantation in the world, which the author visited in 1963 thanks to Indian foresters (Borota, 1965).

An eucalypt plantation was established in 1863 in the Arambyshola forest. Later, a permanent experimental plot was established in this forest within the Nilgiris Forest Division in 1913. At that time the age of the stand was 50 years. The area is situated at an altitude of 2,300 m, and it gently slopes down in a southwestern direction. The climate is uniform and temperate. The average annual rainfall is 1,380 mm. The average monthly temperature ranges from 12 to 17° C. These data refer to the Nilgiris Forest Division. The soil is loamy, deep, moist and rich in humus (Jeyadev, 1954).

The experimental plot has an area of 1.28 ha and supports 73 trees, i.e. 57 trees per hectare. In the lower stratum there are *E. globulus* trees which have come up as coppice after thinning. The plantation was established in 1863 with a spacing of 1.80 by 1.80 m.

All trees on the plot planted in 1863 are marked with numbers in fast colours and their girth at 1.37 m above ground level is pinpointed. Since the plot was established in 1913, periodic measurements have been made, the two most recent ones giving the girth of all numbered trees and the heights of some trees. This was done when the trees were 91 and 99 yeards old. These data were analysed by the author and the results are given in Table 48 and Fig. 100. The frequency distribution of trees is represented, because of the wide variation range of their diameter breast height, by diameter classes at intervals of 10 cm. To obtain a closer characterization of the dynamics of growth, the author used the current annual diameter increment which has been computed from the differences between inventories of all original trees on the plot at the ages of 91 and 99 years.

Table 48. Diameter distribution and current annual diameter increment of trees in *Eucalyptus globulus* plantation. Nilgiris Forest Division

Diameter breast height	Number of trees at the age of		Current annual diameter increment
cm	91 years	99 years	mm
70	1	–	–
80	9	5	4,3
90	19	17	5,0
100	13	16	4,7
110	10	9	6,0
120	8	9	5,2
130	6	7	8,1
140	2	4	6,8
150	2	2	9,6
160	2	2	7,6
170	1	1	13,2
180	–	1	18,2
Sum	73	73	
$\bar{d}$, cm	105.6	111.4	$\Delta\bar{d}$: 7,3

The mean diameter breast height of trees at the time of the establishment of the permanent experimental plot at a stand age of 50 years was $\bar{d} = 72.0$ cm; at the age of 91 years $\bar{d} = 105.6$ cm, and at the age of 99 years $\bar{d} = 111.4$ cm. The standard deviation of the diameter distribution of trees in the stand increased only slightly during the last 8 years, and the value of the variation coefficient as the relative measure of variability remained virtually unchanged during this period. As shown by the computed values of the coefficient of asymmetry, the frequency distribution of trees at stand ages of 91 and 99 years is considerably left-handedly asymmetric ($A = 0.934$; 0.813). The values of the coefficient of excess ($E = 0.432$; 0.337) indicate a moderate pointedness of the frequency distribution of trees compared to the normal distribution.

The current annual diameter increment of trees is computed as the average for 8 years (from the age of the stand between 91 and 99 years). The current annual diameter increment in the eucalypt plantation varied in the range of $\Delta d = 1.0$ mm (trees Nos 30 and 50) to $\Delta d = 18.2$ mm (tree No. 60). Tree No. 60 has the largest diameter breast height = 180.4 cm. Tree No. 50 is one of the smaller diameter trees

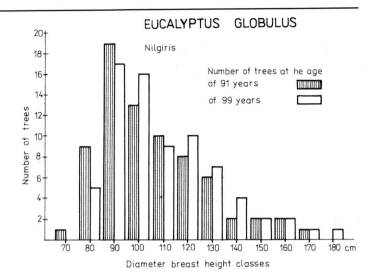

Fig. 100. Diameter distribution of trees in *Eucalyptus globulus*. Nilgiris Forest Division, India.

in the plot, with a diameter breast height = 89.8 cm, but tree No. 30 has a diameter breast height = 128.6 cm.

The current annual diameter increment for 24% of the trees on the experimental plot ranges from 1.0 mm to 4.0 mm, for 42% of the trees it ranges from 4.1 to 6.0 mm, for 23% of the trees it ranges from 6.1 to 10.0 mm, and 11% of the trees have an annual increment of more than 10 mm.

The results of the measurement have revealed that *Eucalyptus globulus* maintains a good ability to increase even at an advanced age. The mean increment sample tree of the plantation has had an annual diameter increment of $\Delta d = 7.3$ mm. The diameter increment is increasing from the thinnest to the thickest tree.

The height of ten trees selected at random was measured using Abney's hypsometer. From the heights measured, the author calculated the average height of the trees on the plot. At a plantation age of 84 years the average height was $\bar{h} = 61.2$ m; at an age of 91 years $\bar{h} = 66.3$ m, and at the age of 99 years $\bar{h} = 72.1$ m. Thus the average height of the plantation increased annualy by some 0.73 m.

Using the approximate equation for computing the crop volume of the plantation:

$$V = BHF,$$

where V is the the cubic volume of the crop, B the total basal area of the crop, H the average crop height, F the average cylindrical form factor for the crop, the current annual crop volume increment of this *Eucalyptus globulus* plantation may be calculated. Basal area and average crop height were obtained from individual tree measurements (the average crop height with the use of sampling). The average crop form factor may present a problem. Assuming the average form factor of the plantation $F = 0.30$, the following crop volumes are obtained: at the age of 91 years $V = 1,272$ m³ (i.e. 992 m³ per hectare), at the age of 99 years $V = 1,539$ m³ (i.r. 1,200 m³ per hectare).

During the last eight years the crop volume of the experimental plot of the *Eucalyptus globulus* plantation in the Nilgiris Hills increased by 267 m³ of timber. The current annual crop volume increment of the experimental plot reaches about 33.4 m³, i.e. 26.0 m³ per hectare.

5.3.2 *Eucalyptus regnans* F. v. MUELL.

Eucalyptus regnans grows in Australia, in the natural forests of Victoria and Tasmania to a diameter of 200 cm and to a height of 90 m. It occurs in plains and hills up to an altitude of 900 m. It does well on rich, loamy soil, at annual precipitation of 750–1,650 mm and an average temperature of the warmest month of 23 °C. The timber is a light to rosy colour, it is medium strong and medium hard and durable. The average specific weight of the dry wood is 0.65 g per cm². It is one of the most widely used timbers in Australia. It is used in construction, for the manufacture of furniture, veneer and plywood and mainly for the production of pulp (Jacobs, 1979).

The introduction of *E. regnans* gave good results mainly in New Zealand, in the Republic of South Africa, in Kenya and Tanzania. In Kenya the plantation of *E. regnans* in Kinango (2,440 m above sea level) astonishes the visitor by its enormous growth; the tallest trees attaining at the age of 35 years a height of 80 m (1982).

In Olmotonyi (northern Tanzania, altitude 1,850 m) the mean diameter breast height of a 13-year-old plantation of *E. regnans* (51 trees) was 24.7 cm ($\Delta \bar{d}$ = 19 mm), and the mean height of dominant trees was 29.8 m ($\Delta \bar{h}$ = 3.7 mm). In the Kigogo Arboretum in southern Tanzania (altitude 2,000 m, annual rainfall 1,800 mm) four trees of *E. regnans* died probably because of drought and were felled at an age of 32 years. Their diameter breast height was 65–89 cm, the tree height was 44–49 m and the volume of stems (with a diameter above 20 cm) was 6.3–9.1 m³ (Borota, 1971a).

In the Lushoto Arboretum in the West Usambara Mountains, Tanzania (altitude of 1,500 m, average annual rainfall 1,070 mm, average annual temperature 18° C) 12 trees of *E. regnans* were measured at the age of 10 years. The mean diameter breast height *(D_G)* was 32.0 cm (D.B.H. = 23.4 to 44.3 cm), and the mean height of these 12 trees was 32.0 m (*H* = 27.5 to 34.0 m) (Borota, 1969).

Among *E. regnans* stand established on smaller plots in Tanzania the greatest rate of increment was observed in a stand having an area of 0.25 ha in the Shagayu Forest Reserve in the West Usambara Mountains. This plantation of *E. regnans* was established in November 1958 with a spacing of 2.7 by 2.7 m on deep black loamy soil. The stand is situated at an elevation of 1,900 m, the average annual rainfall is 1,200 mm.

At a stand age of 8 years a permanent experimental plot for increment observation (experiment No. 670) was established (Fig. 101). All trees growing on the experimental plot which has an area of 0.15 ha were numbered and measured. The

204

Fig. 101. *Eucalyptus regnans,* age: 8 years. Permanent experimental plot in Shagayu Forest Reserve, Tanzania.

second consecutive thinning was carried out in the stand. Statistical characteristics computed from measurements made at the stand age of 8 and 12 years are given in Table 49. Figure 102 shows the graphic representation of trees distribution by height at the same ages together with compensated curves of the normal frequency distribution of trees.

The values of the mean diameter breast height at stand ages of 8 and 12 years indicate a very high diameter differentiation of the stand. In the 8-year-old stand the average annual diameter increment was almost 3 cm and in the 12-year-old stand it was 2.35 cm.

In analysing the frequency distribution of trees by their height and by the mean height of the stand a very important height differentiation of the 8 and 12 years old stands of *E. regnans* is obvious. The height increment of trees was, on average, 3 m annually. The variability of the stand in the tree distribution by height is very small. Whereas in the 8-year-old stand the tree distribution by height was right-hand-edly asymmetric ($A = -0.685$) and more pointed ($E = 0.358$) as compared to the

Table 49. Statistical characteristics of *Eucalyptus regnans* plantation. Experiment 670, Shagayu

Statistical characteristics	Age of the stand	
	8 years	12 years
Number of trees per ha	687	520
Arithmetic mean D.B.H., $\bar{d}$, cm	22.2	28.2
Arithmetic mean annual D.B.H. increment, Δd, mm	27.8	23.5
Standard deviation, s_d, cm	3.11	3.77
Coefficient of variation, v_c, %	14.0	13.4
Coefficient of asymmetry, A	0.110	0.157
Coefficient of excess, E	−0.923	−0.835
Arithmetic mean height, $\bar{h}$, m	26.3	34.6
Arithmetic mean annual height increment, $\Delta \bar{h}$, m	3.3	2.9
Standard deviation, s_d, m	3.11	3.29
Coefficient of variation, v_c, %	14.0	9.5
Coefficient of asymmetry, A	−0.685	0.482
Coefficient of excess, E	0.358	−0.590

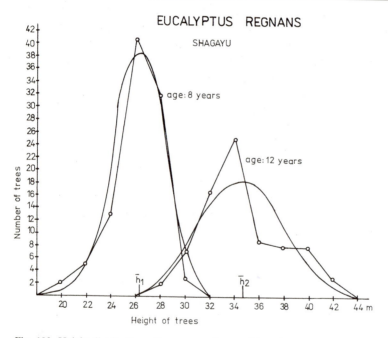

Fig. 102. Height distribution of tree at stand ages of 8 and 12 years. *Eucalyptus regnans*, Shagayu Forest Reserve, Tanzania.

normal distribution, in the 12-year-old stand the frequency distribution of trees is left-handedly asymmetric ($A = 0.482$) and flatter ($E = -0.590$).

The resulting values of the χ^2 test of good agreement in the tree distribution by diameter breast height ($\chi^2 = 4.874$; $\chi^2 = 12.211$) confirm the approximately normal frequency distribution of trees in the stand population.

Timber production on the experimental plot of *E. regnans* in Shagayu is listed in Table 50. The first measurement of trees and the first thinning of the stand were made at a stand age of 6 years. At that time the experimental plot was delimited but the trees were not numbered, and a very strong thinning operation was carried out. Two years later when the stand was 8 years old, 18% of the growing stock was removed from the stand by the second thinning. In the 12-year-old stand no thinning was made.

Table 50. Volume production of *Eucalyptus regnans* experimental plot. Shagayu, Experiment 670

Age of stand in years	Volume production in m³ per ha			Thinning in %
	before thinning	thinning	after thinning	
6	356	148	208	42
8	270	49	221	18
10	282	–	282	–
12	440	–	440	–

The growing stock of the 12-year-old stand of *E. regnans* was very high, equalling 440 m³ per hectare which corresponds to an average annual increment $\Delta V = 36.7$ m³ per hectare. The total timber production of this stand was 637 m³ per hectare, thus the total average annual volume increment was 53.1 m³ per hectare.

From the differences between the two inventories made at the ages of 8 and 12 years, the current annual diameter and height increment of all trees of *E. regnans* on the Shagayu experimental plot was computed and evaluated. The current annual diameter increment of trees varied in the 12-year-old stand from $\Delta d = 4.7$ mm (D.B.H. $= 20.4$ cm) to $\Delta d = 30.2$ mm (D.B.H. $= 36.0$ cm). The current annual diameter increment of the stand was $\Delta \overline{d} = 13.4$ mm. The current annual height increment of the trees on the Shagayu experimental plot varied from $\Delta h = 0.4$ m ($h = 27.7$ m) to $\Delta h = 3.5$ m ($h = 38.5$ m and 42.5 m, respectively). The current annual height increment of the stand was $\Delta \overline{h} = 2.0$ m.

In this very fast growing *Eucalyptus* species the author analysed the tree distribution in the stand at the age of 12 years also as a function of the current annual diameter and height increment. For a clearer understanding the tree distribution in the stand of *E. regnans* is graphically represented as the function of the current annual diameter increment in Fig. 103, and the frequency distribution of trees in the stand as a function of their current annual height increment is shown in Fig. 104. Here the tree distribution with regard to the diameter and height increment of trees has a normal distribution of frequencies. This is quite different from what is found in the natural miombo forests for the species *Afzelia quanzensis* and *Pterocarpus angolensis* (Table 42), where a declining tendency of the diameter increment from the smallest to the largest values of increment was apparent.

The author also analysed the correlation relationship between the diameter breast height *(d)* and the current annual increment of trees *(Δd)* in the eucalypt stand, assuming that there was a linear relationship (Janko, 1947, 1948). Data of the correlation table (Table 51) and the computed value of the correlation coefficient $r = 0.740$ indicate that there is a close relationship between the diameter breast height and the current annual diameter increment of trees in an even-aged eucalypt plantation. They confirm the correlation dependence between the mentioned survey values.

EUCALYPTUS REGNANS

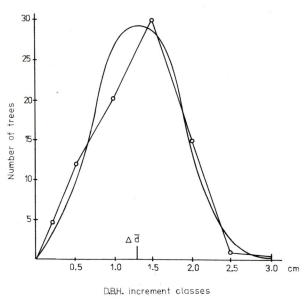

Fig. 103. Tree distribution in a 12-year-old stand as a function of the current annual diameter increment. *Eucalyptus regnans*, Shagayu Forest Reserve, Tanzania.

EUCALYPTUS REGNANS

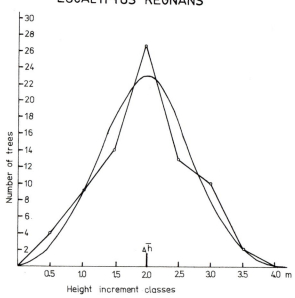

Fig. 104. Tree distribution in a 12-year-old stand as a function of the current annual height increment. *Eucalyptus regnans*, Shagayu Forest Reserve, Tanzania.

208

Table 51. Correlation between diameter breast height and diameter breast height increment of *Eucalyptus regnans*. Shagayu, Experiment 670

d_i $\rightarrow$ / Δd_j $\downarrow$	20	22	24	26	28	30	32	34	36	38	Σn_i	$\bar{d_i}$	$n_i\Delta d_i$	$n_i\Delta d_i^2$	$n_{ij}\Delta d_j d_i$
4	2		1	2	1						6	24.0	24	96	576
8	1	1	7	3	4						16	25.0	128	1,024	3,200
12		1	6	6	4	4	1				22	26.6	264	3,168	7,032
16				3	4	5	3	2			17	29.6	272	4,352	8,064
20				1	2	1	4	6	2		16	32.3	320	6,400	1,0320
24							1			1	2	35.0	48	1,152	1,680
Σn_j	3	2	14	15	15	10	9	8	2	1	79	–	1,056	16,192	30,872
$\Delta \bar{d_j}$	5.3	10.0	9.4	11.5	12.5	14.8	18.2	19.0	20.0	24.0	–				
$n_j d_j$	60	44	336	390	420	300	288	272	72	38	2,220				
$n_j d_j^2$	1,200	968	8,064	10,140	11,760	9,000	9,216	9,248	2,592	1,444	63,632				
$n_{ij}d_i\Delta d_j$	320	440	3,168	4,472	5,264	4,440	5,248	5,168	1,440	912	30,872				

$$\bar{d} = \frac{1}{n}\sum n_j d_j = \frac{2220}{79} = 28.10\text{ cm}$$

$$\overline{\Delta d} = \frac{1}{n}\sum n_i \Delta d_i = \frac{1056}{79} = 13.37\text{ mm}$$

$$s_d = \sqrt{\frac{\sum n_j d_j^2}{n} - \bar{d}^2} = \sqrt{\frac{63632}{79} - 28.10^2} = \sqrt{15.878} = 3.98\text{ cm}$$

$$s_{\Delta d} = \sqrt{\frac{\sum n_i \Delta d_i^2}{n} - \overline{\Delta d}^2} = \sqrt{\frac{16192}{79} - 13.37^2} = \sqrt{26.205} = 5.12\text{ mm}$$

$$r_{d\Delta d} = \frac{\sum n_{ij}d_i\Delta d_j - n\bar{d}\,\overline{\Delta d}}{n s_d s_{\Delta d}} = \frac{1191.063}{1609.830}$$

$$r_{d\Delta d} = 0.740$$

5.4 *Cedrus deodara* Loudon, Pinaceae family

(Synonyms: *Cedrus libani* Barr. var. *deodara* Hook. f.; *Cedrus indica* de Chambr.; *Pinus deodara* Roxb.). Commercial names: deodar, Himalayan cedar. Local names: deodar, dedwar, kelu, kilar.

The natural area of distribution of *Cedrus deodara* is in the mountain regions of the Western Himalayan Mountains. There it grows in the belt of coniferous forests at altitudes from 1,300 to 3,050 m on various geological substrata, on gneiss, granit, mica schist, limestone, slates and other rocks. It prefers sites in valleys and mountain slopes on deep, medium rich soils, mainly on northern exposures (Borota, 1982).

The most consolidated deodar forests occur at altitudes between 1,800 and 2,600 m in India in Jammu and Kashmir, Himachal Pradesh and Uttar Pradesh up to 79°50' E longitude.   \

Air temperature varies in the deodar forest regions from –15 to 35° C. Average annual precipitation is around 850–1,700 mm, the larger part of which falls in the form of snow.

Cedrus deodara mostly forms uneven-aged mixed stands with the coniferous species *Pinus excelsa, Picea smithiana, Abies pindrow, Abies spectabilis, Taxus baccata* and with broad-leaved species such as *Juglans regia, Acer caesium, A. pictum, Aesculus indica, Quercus semecarpifolia, Q. incana, Ulmus villosa, Alnus nepalensis* and *Carpinus viminea* etc. In the undergrowth *Parrotia jaquemontiana, Indigofera gerardiana, Cotoneaster bacillaris, Celtis australis, Viburnum nervosum, Rhododendron arboreum,* species of the *Desmodium* genus and other trees and shrubs occur frequently.

Cedrus deodara grows to large dimensions. The tree with the largest diameter was measured in Mohu Mangat, in the state of Jammu, with a D.B.H. = 376 cm, and the tallest tree was found at Dungri-Kulu, in the state of Punjab, and it measured 73 m. On very good sites it usually grows to a height of 50 m and to a diameter breast height of 150 cm and more. It forms straight, cylindrical boles.

The heartwood is of a yellow to brown colour. It is medium heavy, medium hard, easy to process, and durable. It is used for the construction of buildings and bridges, for railway sleepers and for the manufacture of heavy furniture as well as for wood-carving.

The management of deodar forests

Forest management of the mountainous coniferous forests in India is based on working plans. The deodar forests are included, according to working plans, in working cycles. These are either working cycles based on selective logging or working cycles with approximately even-aged forest stands in the compartment system.

In working cycles based on selective logging, management is concerned with irregular uneven-aged mixed forests. Logging is carried out by felling mature trees

210

one by one or in small groups. The minimum diameter breast height for cutting is usually 60–70 cm and the logging cycle is 25–30 years. According to inventory records, for example, in the Kulu Forest Division (Punjab) the timber stock of mature trees in deodar forests where the working cycle is based on selective logging was 93 m^3 per hectare (Aggarwal, 1949).

In the working cycle with the modified clear cutting system of the compartment system, for example in the Kamraj Forest Division (Kashmir), the deodar forests of the working cycle with an area of about 20,000 ha have been divided into periodical blocks – age classes, each one with an age interval of 30 years. The rotation was set at 150 years. This means that natural uneven-aged stands were gradually converted to approximately unevenly aged stands. Regeneration fellings are carried out in the oldest age class (120–150 years), their aim is the natural regeneration of forest stands (Fig. 105). In the first half of this age class the crown cover is opened up strongly (light cutting or seed felling phase) and in the second half of the age class, after natural regeneration is obtained (usually in groups), the final cutting is finished. In the age class from 90 to 120 years only mature trees, i.e. trees with a diameter breast height exceeding the minimum limit, may be felled. In periodical blocks of the age from 30 to 90 years only thinnings may be carried out, usually removing from the stand trees of inferior quality. The average growing stock of deodar trees with a diameter breast height exceeding 30 cm in the working cycle of the compartment system was 180 m^3 per hectare in the Kamraj Forest Division (Kashmir) (Kaul and Fotidar, 1961).

Deodar forest stands are also established artificially by sowing or planting. Because the seed quickly looses its germination capacity, it is sown in the forest terrain or in nurseries in autumn immediately after becoming mature. For planting, strong three-year-old plants 30–40 cm tall are used.

Growth and increment of deodar forests

In order to obtain information on the growth and increment of trees and stands in deodar forests, permanent experimental plots were set up at the beginning of this century. On such plots the average age of trees was determined and the diameter breast height and height of randomly selected trees of average growth were measured.

Based on 83 such measurements on permanent experimental plots in the main deodar regions of the western Himalayas, preliminary yield tables for approximately even-aged deodar stands were set up in India at the end of World War I (Troup, 1921). Estimation values of these yield tables for the first site class converted to the metric system are given in Table 52. These preliminary tables did not contain data on thinnings. The growing stock was computed, starting from a diameter breast height of 20 cm, for felled sample trees measured by section. The average annual volume increment of the stand (9.1 m^3 per hectare) culminated on the first site class

Fig. 105. *Cedrus deodara.* Natural regeneration in the Lolab Range, Kashmir, India.

Fig. 106. *Cedrus deodara* plantation, aged 85 years. Manali Forest Reserve, Punjab, India.

Table 52. *Cedrus deodara.* First provisional yield tables in India

Age in years	Stem number	D.B.H.	Mean height	Stand volume	Mean stem volume
	ha	cm	m	m³ per ha	m³
1st quality class					
20	2,347	11.3	8.5	–	–
30	1,507	17.7	14.3	69	0.05
40	988	24.2	19.5	193	0.19
50	692	31.5	24.0	310	0.45
60	494	38.8	26.8	435	0.88
70	395	45.3	32.3	552	1.40
80	321	51.7	36.0	662	2.06
90	272	58.2	38.4	780	2.87
100	247	63.8	40.8	890	3.60
110	222	69.5	42.7	987	4.45
120	210	74.4	43.9	1,097	5.22
130	198	79.2	44.8	1,180	5.96
140	185	84.1	45.7	1,242	6.71
150	173	88.1	46.6	1,297	7.50

at the age of 120 years. The trees grow to a diameter of 80 cm by the age of 130 years and produce annually on average 6 m³ of stem timber per hectare.

The Kulu Forest Division in Punjab is one of the most important regions for both natural deodar forest and artificially established deodar stands (Aggarwal, 1949) (Fig. 106). There is in this region a number of permanent experimental plots in maturing deodar stands, unfortunately with rather a small area (0.08–0.20 ha). One of these plots, No. 3, aged 85 years and situated in the Bias river valley at an altitude of 1,800 m is characterized by the following data:

area	0.17 ha,
number of trees	71, i.e. 422 per hectare,
basal area	17.3 m^2, i.e. 102 m^2 per hectare,
mean diameter breast height	55.6 cm,
diameter range	33.0–68.6 cm,
mean height of the stand	36.0 m,

According to empirically obtained numerical data, the stand can be classified as belonging to the first site class.

More detailed data obtained from measurements taken in the permanent experimental plot No. 21, likewise in the Kulu Forest Division, were analysed in more detail. The experimental plot was established in 1931 at a stand age of 50 years. The stand was established artificially. It is situated in the Bias river valley at an altitude of 1,450 m. The soil is loamy to gravelly and medium rich in nutrients. The average annual precipitation is 1,300 mm in this region and the air temperature drops to –10° C in winter, in summer it reaches 33° C. The area of the plot is 0.15 ha.

At a stand age of 50 years on the permanent experimental plot No. 21 there were 357 trees per hectare and there was a total of 487 m^3 per hectare of stem timber (D.B.H. $\geq$ 20 cm). In the observation period of 21 years, 17% of the trees and 14% of the stem timber stock were removed from the stand by thinnings. At an age of 71 years there were 295 trees per hectare on the experimental plot. The mean diameter breast height of the stand was 53.8 cm, the mean height of the stand was 34.1 m and the stem timber stock was 747 m^3 per hectare. The fully stocked stand may be classified as belonging to the first site class. The total current annual volume increment of the stand for the observed 21-year period was 15.7 m^3 per hectare.

The author used the numerical data from the inventories of the stand at the ages of 50 and 55 years, and 66 and 71 years for an analysis of the diameter distribution of trees and for an analysis of the diameter increment. The following basic statistical characteristics were computed: arithmetical mean diameter $(\bar{d})$, standard deviation (s_d) and variation coefficient of the diameter distribution of trees (v_c); they are given in Table 53. There were 44 trees on the experimental plot at a stand age of 50 and 55 years, and 43 trees at the age of 66 and 71 years.

The values of the arithmetical mean diameter indicate a very good diameter development of the deodar stand. The standard deviation increases with greater age. The variation coefficient (as the relative measure of variability) indicates a relatively small variability of the diameter distribution of trees, which has a declining tendency with increasing age.

The computed values of the annual diameter increment, of its standard deviation, and the variation coefficients are also given in Table 53. The current annual diameter increment is computed from the difference of two inventories of all trees on the plot at the stand ages of 55 and 71 years. The annual increment of individual trees is taken as the average of five years (from the age of 50 to 55 years, and 66 to 71

214

Table 53. *Cedrus deodara.* Data on diameter and diameter increment from the permanent experimental plot No. 21. Forest Division Kulu

Age	Stem number	$\bar{d}$	s_d	v_c	$\overline{\Delta d}$	$s_{\Delta d}$	$v_{\Delta c}$
		cm	cm	%	mm	mm	%
50	44	41.5	7.58	18.3	5.8	2.60	51.1
55	44	44.4	7.55	17.0			
66	43	50.0	8.63	17.3	5.6	2.06	36.6
71	43	52.8	8.44	16.0			

years, respectively). The current annual diameter increment for the whole plot expresses the total of the annual diameter increments of all trees divided by the number of trees in the experimental plot.

Cedrus deodara is of very good growth in the observed region. The current annual diameter increment of the deodar stand on experimental plot No. 21 may be characterized as being very good ($\overline{\Delta d} = 5.6$–$5.8$ mm). The variability of the current annual diameter increment declines with the increasing age of the stand ($s_{\Delta d} = 2.60$–2.06 mm). The values of the variation coefficient indicate a greater variability in the distribution of the diameter increment, than that which is found in the distribution of trees by their diameter breast height.

Periodical measurements of growth values on permanent experimental plots are a source of valuable material for the study of growth potential of this most valuable coniferous tree species of the Himalayas. Evaluated dendrometrical data served as a basis for the construction of stand volume and yield tables.

5.5 *Juniperus procera* Hochst. ex Endl., Pinaceae family

Juniperus procera, the East African cedar is the best-known autochthonous coniferous tree species in eastern Africa. It grows to heights of 35–40 m and to diameters breast height of 80–120 cm. It occurs on various soils in mountainous regions from Ethiopia down to northern Tanzania. It grows at altitudes of 1,500–2,700 m and where the average annual precipitation is between 1,100–1,400 mm.

The timber is of a russet colour and it is medium hard and heavy, easy to work, and durable. It is of wide use. For good quality timber the species is also cultivated in plantations.

The following survey data were obtained from one of the oldest plantations of *Juniperus procera* in Tanzania which was established in 1909 near Lushoto (Western Usambara Mountains), i.e. experimental plot No. 66 (Fig. 107) which was 61 years old:

Fig. 107. *Juniperus procera*, aged 60 years. Lushoto Forest District, Tanzania.

area of the experimental plot – ha	0.55,
number of trees per hectare	182,
mean diameter breast height of the stand – cm	47.1,
mean annual diameter increment – mm	7.7,
mean height of the stand – m	32.5,
average annual height increment – m	0.53,
basal area of the stand – m^2 per hectare	32.0,
growing stock – m^3	347,
average annual volume increment – m^3 per hectare	5.7,
volume of the mean stem of the stand – m^3	1.91.

The experimental plot is situated at an altitude of 1,450 m. The average annual rainfall is 1,070 mm and the average annual temperature is 18° C.

The tree distribution by diameter breast height in the 61-year-old stand has a normal frequency distribution according to Laplace–Gauss.

The author carried out an inventory of plantations of *Juniperus procera* which are situated in the Western Usambara Mountains in the Shume and Magamba

Forest Reserves (Table 54). The area of the plantations was 336 ha. The inventory was made on sample plots. Timber cruising was carried out on 2.46% of the area of all plantations, beginning from a diameter breast height of 10 cm (Borota, 1966). Stands belonging to the 41–50 years age class were of the poorest quality and were the least productive.

Table 54. The results of the inventory of *Juniperus procera* plantations in Tanzania

Age classes in years	Area		Number of trees	Mean D.B.H.	Basal area	Growing stock
	ha	%	ha	cm	m² per ha	m³ per ha
1–10	–	–	–	–	–	–
11–20	13.4	4.0	1,396	13.2	19.0	60
21–30	178.3	53.1	919	21.1	31.3	198
31–40	54.6	16.2	640	25.7	31.8	294
41–50	89.4	26.6	546	23.9	21.9	186
51–60	0.4	0.1	319	37.6	35.6	498
Total	336.1	100.0	–	–	28.4	205

The mean diameter breast height and the mean height of the various plantations varied with regard to age and especially with regard to the sites of the stands. The mean height of the stand fluctuated in the age classes from $\overline{h} = 8.8$ m at the age of 16 years to 30.5 m at the age of 60 years. The tree distribution by diameter breast height in the various age classes corresponded to the normal frequency distribution. The range of variation within the stand populations increased regularly with the increasing age of the stand.

5.6 *Pinus caribaea* MORELET, **Pinaceae family**

Pinus caribaea, the Caribbean pitch pine at present is the most frequently cultivated pine in the tropics. It does well when introduced on sandy, sandy to loamy and loamy, well-drained soils in coastal lowlands. It is also planted in mountain regions up to altitudes of 1,700 m where there is 800–2,000 mm of rainfall annually. It grows to a diameter of 60–80 cm and to a height of 30–40 m. The timber of the Caribbean pine is strong, heavy, of good quality and it contains much resin. It is used mainly in construction, for posts and is suitable for the production of pulp (Burley and Nikles, 1972, 1973).

From a practical point of view *Pinus caribaea* is divided into three varieties (Lamb, 1973):
1. *Pinus caribaea* var. *bahamensis:* it grows naturally on the Bahama Islands (24–27° N). In its area of natural distribution this variety attains a height of 15–20 m.

217

2. *Pinus caribaea* var. *caribaea* occurs naturally in western Cuba and on the Isle of Pines (22–23° N). It grows to a height of 20–27 m.
3. *Pinus caribaea* var. *hondurensis* is naturally distributed along the shores of the Atlantic Ocean in Central America from Belize to northern Nicaragua (12–18° N). It grows to a height of 35 to 40 m and produces annually 10–40 m³ of timber per hectare.

On the oldest experimental plot established at Imbile in 1930 in Queensland, Australia (31° S, average precipitation 1,140 mm), pine trees of *P. caribaea* var. *caribaea* attained a diameter of 40 cm and a height of 30 m at the age of 35 years.

In Tanzania *P. caribaea* has been cultivated since 1953. In 1971 there were in the country over 80 experimental plots of Caribbean pine. About 2,750 ha of plantations existed along the sea coast (80–700 m above sea level) (Fig. 108). In the Victoria Lake region (1,150 m above sea level), and in the Ukaguru Mountains (1,000–1,150 m above sea level), most plantations were younger than 5 years. Trial plots have been established in a wide variety of sites in various parts of the country. The seed of this pine originated mainly in Belize but later on it also came from the Bahamas and from Guatemala.

The author analysed the growth of Caribbean pine on experimental plots in Tanzania and published the results (Borota, 1972) (Table 55, where the individual plots are listed in order of age, from the oldest to the youngest crops). The experimental plots with *Pinus caribaea* were established at an altitude of 20–3,200 m on different types of soil with average annual precipitation of 800–1,800 mm. The number of trees on the plots at the time of their measurement varied from 27 to 400. The average annual diameter increment on several plots exceeded 2 cm and the average annual height increment was over 2 m. The largest total timber production of Caribbean pine was observed on the experimental plot in the Busenge Forest Reserve, near Lake Victoria. Here at a stand age of 8.5 years, 490 m³ of timber was recorded per hectare (2,000 trees per hectare) (Fig. 109).

On Maesome Island in Lake Victoria (1,160 m above sea level, average annual precipitation of 1,300 mm) in a stand aged 8.5 years, 1,833 trees per hectare and a growing stock of 363 m³ per hectare were established. The average annual volume increment was 42.7 m³ per hectare. The average diameter breast height of the stand was 19.7 cm and the top height 16 m. This stand has been established on deep, yellowish-brown, permeable loamy soil in a spacing of 2.4 by 2.4 m. Seed was obtained from the then British Honduras (now known as Belize). The original vegetation was a natural high forest of limited productivity.

The arithmetical mean diameter breast height of the stand $\bar{d} = 19.0$ cm confirms its very high diameter development. The average annual diameter increment in this 8.5-year-old pine stand was $\Delta \bar{d} = 22.4$ mm. The standard deviation of the stand population $s_d = 3.9$ cm and the value of the variation coefficient $v_c = 20.5\%$ indicate no great variability of tree diameters. The normal frequency distribution of trees of this stand population is confirmed by the resulting value $\chi^2 = 6.58$ which for eight degrees of freedom does not exceed its critical value ($\chi^2 = 15.5$).

218

Fig. 108. *Pinus caribaea* growing on poor sandy soil. Aged 5 years, average diameter breast height: 10 cm, average height: 8 m. Region of Lindi, Tanzania.

Table 55. The growth-rate of various experimental plots of *Pinus caribaea* in Tanzania

Order	Experiment	Locality	Altitude	Rainfall per year	Age	Stems	D_G	MDI	DH	DHI	Basal area	Volume	MAVI
			m	mm	years	per ha		cm		m	m² per ha	m³ per ha	m³ per ha
1	70-1	Kitulo plateau	3,200	1,450	17.6	1,630	10.2	0.58	7.5	0.43	13.4	42	2.4
2	70-3	Kitulo plateau	2,680	1,450	17.6	1,440	16.4	0.93	12.2	0.69	30.5	151	8.6
3	103-5	Njombe-Lwango	1,740	1,100	15.7	1,650	25.4	1.62	18.8	1.20	83.4	567	36.1
4	103-6	Njombe-Mswima	1,700	1,100	15.7	721	31.3	1.99	19.1	1.27	55.4	388	24.7
5	103-7	Njombe-Kidegembe	1,560	1,050	15.7	692	33.6	2.14	22.0	1.46	61.5	492	31.3
6	ARB-42	Lushoto	1,500	1,050	14.5	670	33.9	2.34	26.7	1.84	55.6	600	41.4
7	329-2	Ruiga River	1,150	1,130	11.8	1,100	20.2	1.71	18.5	1.56	35.1	253	21.4
8	675	Geita-Vuziligembe	1,200	800	11.8	912	21.8	1.84	18.2	1.54	34.0	252	21.4
9	576	Rubya-Ukerewe	1,140	1,700	11.8	844	29.2	2.48	22.5	1.91	56.4	490	41.5
10	288-5	Rubya-Ukerewe	1,130	1,700	11.6	1,170	24.3	2.09	20.7	1.78	54.7	473	40.7
11	356	Minziro-Bukoba	1,160	850	11.5	1,082	18.3	1.59	16.6	1.44	28.5	192	16.7
12	87	Rondo plateau	750	1,200	11.4	1,080	21.3	1.87	17.1	1.50	38.4	259	22.7
13	630	Pugu-Dar es Salaam	280	1,000	11.3	1,010	19.3	1.70	16.8	1.48	29.7	207	18.3
14	490 A	Ruasina-Bukoba	1,150	1,350	10.0	1,205	20.6	2.06	20.2	2.02	40.2	318	31.8
15	622	Rubare-Bukoba	1,200	1,800	10.0	1,010	26.6	2.66	21.0	2.10	56.3	492	49.2
16	482	Ruiga River	1,170	1,150	8.7	1,940	19.9	2.03	18.0	1.65	60.3	430	49.4*
17	747	Busenge-Lake R.	1,150	1,200	8.5	2,000	21.5	2.52	16.9	1.98	72.7	494	58.1*
18	535-1	Maisome Island	1,160	1,300	8.5	1,833	19.7	2.32	16.0	1.88	55.7	363	42.7*
19	510	Kome Island	1,150	1,200	8.0	1,200	22.1	2.76	17.2	2.15	46.1	330	41.2*
20	535-7	Maisome Island	1,160	1,300	7.8	1,750	19.6	2.51	15.8	2.03	53.1	344	44.1*
21	488	Ruiga River	1,180	1,150	7.6	1,900	18.5	2.43	16.0	2.10	38.2	260	34.2*
22	307	Longuza-Tanga	200	1,600	7.4	1,980	16.3	2.20	16.2	2.19	41.6	280	37.8*
23	ARB-134	Lushoto	1,500	1,050	7.3	1,710	16.0	2.19	12.2	1.69	34.4	162	22.2
24	668	Kigwa-Tabora	1,200	880	6.6	1,562	11.8	1.78	8.0	1.21	17.3	55	8.3
25	669	Urumwa-Tabora	1,220	890	6.1	2,042	13.8	2.26	12.2	2.00	30.4	137	22.4*
26	288 D	Rubya-Ukerewe	1,130	1,700	5.9	1,664	15.7	2.66	11.9	2.01	32.3	145	24.6*
27	550 J	Pugu-Dar es Salaam	280	1,000	5.6	2,125	14.7	2.63	10.8	1.93	36.1	146	26.1*
28	432-2	Uzigwa-Bagamoyo	350	900	5.1	2,140	7.0	1.37	8.1	1.60	8.2	27	5.4*
29	432-4	Uzigwa-Bagamoyo	350	900	5.1	2,000	7.5	1.47	7.3	1.43	9.0	28	5.6*
30	696-2B	Bana-Dar es Salaam	80	1,100	5.0	1,542	14.1	2.82	11.0	2.20	23.9	105	21.0*

* No thinning was done.

Fig. 109. *Pinus caribaea*, aged 8 years. Busenge Forest Reserve, Tanzania.

A high timber production was also observed on the experimental plots of Caribbean pine in Rubare and Rubya Forest Reserves. The experimental plot No. 622 in Rubare was 10 years old, it lies near the town of Bukoba, west of Lake Victoria. On the experimental plot there were 1,010 trees per hectare with a growing stock of 492 m^3 per hectare. The average annual volume increment was 49.2 m^3 per hectare and the volume of the mean stem was 0.49 m^3. The mean diameter breast height of the stand was 26.6 cm (D.B.H. = 12.5 to 38.7 cm) and the mean height of the stand was 19.0 m (h = 16.0 to 23.5 m). The basal area of the stand was 56.3 m^2 per hectare. Being too dense, the stand required urgent thinning. This removed from the 10-year-old stand 180 trees per hectare, i.e. 18% of all trees, with a timber volume of 77 m^3 per hectare, or 16% of the growing stock. The stand has been established on rich loamy soil at an altitude of 1,200 m with an annual rainfall of 1,800 mm.

Empirical numerical data of the diameter distribution of trees, and their theoretical values for the normal frequency distribution of trees on the experimental plot (No. 622) of *P. caribaea* in Rubare Forest Reserve are represented graphically in Fig. 110. This graphic representation indicates a normal frequency distribution of trees by their diameter breast height. This assumption is confirmed by the resulting value of the χ^2 test of good agreement for 13 degrees of freedom (χ^2 = 12.496) which does not exceed its critical value (χ^2 = 22.4). From the computed statistical characteristics, the value of the arithmetic mean diameter $\overline{d}$ = 25.6 cm draws attention to the very high diameter development of the stand. The average annual diameter increment was $\Delta \overline{d}$ = 25.6 mm. The standard deviation s_d = 5.47 cm (as the measure

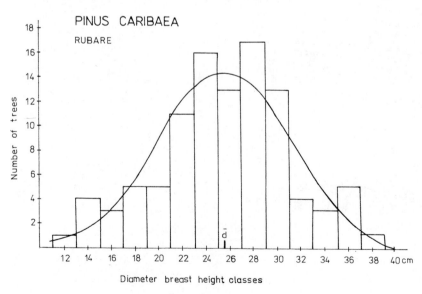

Fig. 110. Empirical and theoretical values of the normal tree distribution as the function of the diameter breast height. *Pinus caribaea,* aged 10 years. Rubare Forest Reserve, Tanzania.

222

Fig. 111. *Pinus caribaea,* aged 10 years. Rubya Forest Reserve, Tanzania.

of variability), and the variation coefficient $v_c = 21.38\%$ (as the relative measure of variability), indicate an approximately equal variability of the diameter distribution of trees as was found in the previously experimental plot.

The experimental plot No. 576 is situated in Rubya Forest Reserve (Fig. 111) also to west of Lake Victoria. At a stand age of 12 years there were 840 trees per hectare with a growing stock of 490 m³ per hectare. The mean diameter breast height of the stand was 29.2 cm and dominant height was 22.5 m. The volume of the mean stem was 0.58 m³ and the average annual volume increment of the stand was 41.5 m³ per hectare.

Pinus caribaea has proved reasonably hardy under a wide variety of conditions in Tanzania. Occasional trees have died out. Giant rats have caused damages on the Coastal Afforestation Project probably because the weeding was not sufficient. In some localities, chiefly on old forest sites the trees are attacked by *Armillaria mellea*. The trees have died in drier sites as a result of drought (Mombo Arboretum). This species is partly resistant to scorching by fire due to its thick bark.

5.7 *Pinus patula* SCHIEDE et DEPPE in SCHLECHT. et CHAM.

Pinus patula is a tree species of the subtropics. It is naturally distributed in central and eastern Mexico in the Tamaulipas, Querétare, Hidalgo, Puebla, Vera Cruz and Oaxaca states.

P. patula forms pure or mixed stands on deep, moist, permeable, loamy or sandy to loamy soils at altitudes between 1,800 and 2,750 m (sporadically up to 3,050 m) with an average annual temperature of 12–20° C. The annual amount of rainfall varies from 1,000 to 1,500 mm; 90% of the precipitation occurs from May to October (Wormald, 1975). Mixed stands are formed by *P. patula* on drier sites with *Pinus teocote* and *P. greggii;* on moister sites it grows together with *P. rudis* and *P. montezumae*. In central parts of Mexico *P. patula* also forms mixed stands with *P. leiophylla, P. pseudostrobus, P. michoacana* and with other species.

In its natural area of distribution *P. patula* attains a height of 10–25 m and sporadically on deep soil reaches height of 30–40 m, and a diameter of 60–120 cm (Vazquez, 1962). It usually forms straight cylindrical boles.

The timber is of a light yellow to light brown colour. It is light, soft, fragile and easy to work. It contains 1–2% of resin and the average specific weight of the wood is 0.45. In Mexico the timber of *P. patula* is used in light construction, carpentry, for packing, parquetry and as fuelwood.

For its very good growth potential, adaptability and relatively good resistance to diseases, *P. patula* is cultivated as an introduced species in plantations, and also as a decorative tree in many tropical and subtropical countries, mainly in southern and eastern Africa (Streets, 1962).

As an introduced species it grows well in regions with clear-cut rainy and dry seasons, with an annual precipitation ranging from 750 to 2,000 mm, and average monthly temperatures of between 8 and 26° C. It does extremely well on deep fertile soils which also conserve moisture in the dry season. It also has a good production potential at altitudes from 1,000 to 2,700 m (Wormald, 1975).

P. patula is cultivated in plantations mainly in countries to the south of the equator; in South Africa to the south of 31° S and in New Zealand as far south as

42° S. There are at present in Africa about 500,000 ha of plantations of *P. patula*. According to accessible data (Wormald, 1975), in 1972 there were in the Republic of South Africa 235,000 ha of plantations of *P. patula* which represents 47% of the area of all coniferous plantations in the country. About three quarters of the timber production of this tree species are used in sawmilling, the remainder is used for pulp production. The rotation is 30–35 years for sawn wood production and 15–20 years for pulpwood production.

In Swaziland there are about 50,000 ha of *P. patula* plantations. About 40% of the produced timber goes into sawmilling, and 60% is used for pulp.

On Madagascar in 1980 there were about 112,000 ha of coniferous plantations (Direction des Eaux et Forêts et de la Conservation des Sols R. D. Madagascar, Antananarive, 1981) including *P. patula* cultivated on an area of about 70,000 ha.

In Zimbabwe *P. patula* has been cultivated since 1920. There are about 40,000 ha of plantations of this tree in the country. It is cultivated in mountain regions where the annual precipitation is above 900 mm.

In Kenya plantations of *P. patula* exist over an area of about 40,000 ha. About 3,000 ha of land is being afforested each year at present. Afforestation is concentrated mainly on mountain savannas and in degraded mountain forests after their cutting (Václav, 1983).

In Tanzania there are about 25,000 ha of *P. patula* plantations. Data on the growth of this tree species on certain localities in Tanzania are given in Table 56 (Fig. 112).

Table 56. The growth of *Pinus patula* on various sites in Tanzania

Location	Altitude	Age	Mean D.B.H.	Mean height
	m	years	cm	m
Ukaguru	1,400	9	16.6	15.3
Sao Hill	2,050	10	16.1	12.8
Kamanga	1,670	13	19.0	12.2
Lushoto	1,500	15	24.0	23.1
Kiwira	2,350	16	32.8	23.1
Kitulo	2,550	18	30.6	21.5
Old Moshi	1,750	25	39.8	28.2
Uru	1,830	25	38.3	26.5
Rongai	2,100	25	34.4	29.7
Kigogo	1,800	35	35.9	28.3

Interesting information is available from the oldest tree increment plots laid down in Kilimanjaro area, localities: Old Moshi, Uru and Rongai which are situated at an altitude of 1,750–2,100 m. In an age of the stands of 25 years, the mean diameter breast height varied from 34 to 40 cm and mean height from 26.5 m to 29.7 m. In a stand age 15 years at Lushoto Arboretum in an altitude of 1,480 m with the average

Fig. 112. *Pinus patula* plantations, Kiwira Forest Reserve, the Rungwe Mountains in the background, Tanzania.

Fig. 113. A 25-year-old stand of *Pinus patula*. Mufindi Region of Sao Hill, Tanzania.

226

annual rainfall of 1,070 mm, the mean diameter breast height of the stand was 24 cm and mean height 23 m (Borota, 1971c).

In a 25-year-old stand of *P. patula* in Kigogo, Tanzania (Fig. 113) the total timber production of a heavily thinned stand was 545 m^3 per hectare, the mean diameter breast height was 34 cm and top height was 25 m. There were 300 trees per hectare (Borota, 1971b).

The area of *P. patula* plantations established in mountain regions of Latin America has so far been smaller, representing about 75,000 ha; most of this area, about 25,000 ha, is in Brazil. Other plantations are in countries of southeastern Asia and in the Pacific.

Timber production in artificially established *P. patula* stands is high. Twenty-year-old stands produce 200–400 m^3 of timber per hectare. At an age of 30 years the growing stock of the main stand is usually around 400–550 m^3 per hectare. For example in a 42-year-old stands of *P. patula* in New Zealand (Whakerenwarewa, according to Weston, cited in Wormald, 1975) 380 m^3 per hectare was removed from the stand by thinnings from a total timber production of 970 m^3 per hectare. The growing stock of the main stand was 590 m^3 per hectare with 270 trees per hectare. The mean diameter of the stand (D.B.H.) was 44 cm and the top height was 32 m.

Such forest plantations, as described above, are usually established in many developing countries on land unsuitable for agricultural crops, on mountain savannas, or as a replacement for forests of low productivity.

Pinus patula as an exotic species has mostly been planted in eastern Africa into higher elevation in highland sites, where there is a rainfall of 900 mm and higher. It has grown remarkably well, the form of the trees is good. It is susceptible to wind damage; it has a reputation for being drought sensitive and very sensitive to fire damage. *P. patula* is a thin barked tree. The yield of *P. patula* plantations varied from 10 to 30 m^3 per hectare annually on a 35-year rotation at a number of 200–350 per hectare (Wormald, 1975).

6 Thinnings in *Pinus patula* stands in Tanzania

From the very beginning of organized forestry, foresters, practicians and theoreticians have studied the problem of tending interventions in forest stands. This has been mainly with regard to thinnings in order to increase, as far as possible, the timber production of forest stands.

The problem of thinnings has been so far one of the most difficult and, at the same time, economically most important problems of forest management. To this day, foresters opinions differ as to the optimum method and intensity of thinning. Although many forest research institutes have gathered and evaluated enough information based on long-term thinning research in forest stands, it is impossible to foretell with certainty, how a given stand will react to a particular thinning intervention.

In marking thinnings, foresters are influenced by certain opinions and hypotheses. Although any forester, in marking thinnings, agrees with the selection of a certain thinning method, each one of them marks the trees in the forest stand for thinning according to his own individual ideas. In any case the forester intervenes periodically, by repeated thinnings in the growth process of forest stands. The thinning interventions in the forest stand influence the timber producing components of the stand. These are namely: the number of trees, the diameter distribution of trees, the mean diameter breast height, the basal area, the mean height of the stand and the increment of both the individual trees and the whole stand. Thinnings are part of the total timber production. This includes the actual volume increment formed by the trees of the main stand and that formed by the trees which are removed by thinning; thus diminishing the growing stock at the time of thinning.

According to results of long-term research, no total increase of timber production may be expected by thinning. The main purpose of thinning is the improvement of the quality of the stand, i.e. the promotion of the increment of trees of the best quality, the improvement of resistance mainly against wind damage, and the improvement of other environmental benefits derived from the forest stand.

In countries of the temperate climatic zone thinning instructions are based on the generally recognized rule that thinnings should be made in forest stands to such an extent as to initially fulfil their tending task. This includes: the improvement of the growing environment and the sanitary conditions of the stand; increasing the stand's resistance to damaging agents, and improving the quality of timber, so as to obtain the maximum possible timber production from the unit area. This means that thinning method should be adjusted to the tree species, to the selected silvicultural system and to the purpose, age and state of the stand.

The situation in the tropical and subtropical regions is more complicated, since here the growth process of trees is much faster than in the temperate zone. Here it is also necessary to select (according to the tree species) the method and the intensity of thinning which would comply with the silvicultural requirements of the stand in question.

Pinus patula is among the most widely distributed introduced pine species in the countries of eastern Africa. In Tanzania until recently there was no experience with thinning fast growing introduced coniferous tree species, therefore modified thinning instructions based on practical experience of foresters in southern Africa were used (Craib, 1939). These instructions were based on the constant number of trees as a function of age on the area unit (Borota and Procter, 1967). Practical experience showed that there was a need for the establishment of the country's own thinning experiments (Willan, 1964).

The oldest research of thinning of a *Pinus patula* stand started in Kigogo in 1951 on two experimental plots. Thinning was carried out on both plots altogether five times. The first thinning was made when the stand was 7 years old and the last one was done when it was 20 years old. Inventories were made on the plots altogether seven times, the last time at a stand age of 25 years.

Thinnings on both plots were on the whole very heavy. At the age of 20 years only 5% of the original trees remained on plot A and 12% remained on plot B (Borota, 1971b).

In the nineteen-sixties further thinning research was established in the countries of eastern Africa (including Tanzania) in stands of *Pinus patula*. An analysis of results of two research programmes are given below.

6.1 Thinning experimental plots in Kiwira

6.1.1 Methodology

Thinning research in the artificially established stand of *Pinus patula* in Kiwira on the experimental plot No. 345 has been based on the constant number of trees remaining after the last thinning until the end of the rotation (Parry, 1955).

The stand was planted in January 1960 with a spacing of 2.7 by 2.1 m, in the region of mountain savannas at an altitude of 2,200 m. The soil is deep, grey-brown and of volcanic origin. In the region of Kiwira, annual precipitation is around 1,350 mm; the rainy season lasts from December to the beginning of May.

The thinning plots were permanently delimited on the ground and all trees in them were numbered and measured at a stand age of 2.5 years. Also in this year the first thinnings were carried out.

The thinning plots formed two parallelly marked series, using the clinal layout of the so-called wedge-shaped areas (Pudden, 1955). One series consisted of plots numbered 1–10, and the other, separated by a 15 metre wide isolation belt, contained the plots numbered 11–20 which were numbered in the opposite direction (Fig. 114). The figure also shows the constant number of trees per hectare after the last thinning. Each thinning plot with an unequal number of trees is wedged into the following plot.

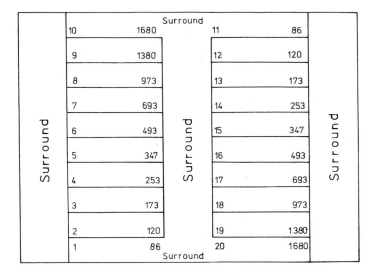

Fig. 114. Pattern of the distribution of thinning plots, experiment 345 Kiwira Forest Reserve, Tanzania.

The area of each thinning plot was 0.075 ha, so that two plots together measured 0.15 ha. In the first thinning year all trees on the thinning plots and on the isolation belts were pruned approximately to one half of the tree height.

The assigned intensity of thinning and the thus reduced number of trees per hectare on the mentioned plots are given in Table 57. Plots 1 and 11 with the strongest thinning and plots 10 and 20 without any thinning were not measured, they acted as isolation belts.

By thinning, trees of lower quality and poorer growth were removed from the stand. The uniform distribution of trees on the thinning plots was also taken into account.

Taking the first inventory data from 1962 as an example, the author analysed the dispersion of tree diameters breast height between the various thinning plots and between the existing pairs of thinning plots. For testing, the F test at a level of 95% significance was used.

By carefully measuring thinning plots annually, it was possible to evaluate their management using mathematical statistical methods.

230

Table 57. *Pinus patula*, Experiment 345. Thinning scheme in the stands at Kiwira

Thinned plots	Final stocking		Age for thinning to stems per ha								
number	per ha	%	1,384	988	692	494	346	247	173	123	86
1 + 11	86	6		3	4	5	6	7	8	9	10
2 + 12	123	9		3	4	5	6	7	8	9	
3 + 13	173	13		3	4	5	6	7	8		
4 + 14	247	18		3	4	5	6	7			
5 + 15	363	25		3	4	5	6				
6 + 16	494	36		3	4	5					
7 + 17	692	50		3	4						
8 + 18	988	71		3							
9 + 19	1,384	100	3								
10 + 20	1,680	121				not thinned					
Approx. year of thinning			1962	1962	1963	1964	1965	1966	1967	1968	1969

6.1.2 Analysis of results

Results of the analysis of the dispersion of tree diameters on the thinning plots showed that at the beginning of thinning research, at a stand age of 2.5 years, the differences between dispersions were not significant in most cases (Jeffers, 1960). Significant differences of dispersion were found between the pair of thinning plots numbered 3 and 13.

For a better understanding, Table 58 gives the distribution of trees as the function of their diameter at the height of 1.3 m above ground. This is given together with the main statistical characteristics of the diameter distribution of trees on the recorded thinning plots before marking the thinnings at a stand age of 2.5 years. Data in Table 58 indicate a homogeneous stand for all pairs of thinning plots with the exception of the pair of plots 3 + 13 (where mainly the numerical values for the standard deviation and of the variation coefficient, taken globally for the whole stand, are recognizably higher).

The studied pairs of thinning plots were also subjected by the author to the analysis of dispersion after the last inventory at a stand age of 13.5 years. The differences of dispersion between the various pairs of thinning plots were, from the statistical point of view, insignificant.

The normal distribution of tree numbers in various years of stand inventory has a regular course on all thinning plots, be it a population of trees before, or after thinning, or be it a population of trees cut in the thinning.

For a clearer idea, the diameter distribution of trees of some pairs of thinning plots is represented graphically. The empirically obtained numerical data are represented in Fig. 115 by histograms, and the theoretical distribution by frequency

Table 58. *Pinus patula,* Experiment 345. The distribution of the number of trees as a function of diameter breast height and the main statistical characteristics in the stands to be thinned prior to marking the trees for thinning at an age of 2.5 years

D.B.H.	Thinned plots								Total
cm	2 + 12	3 + 13	4 + 14	5 + 15	6 + 16	7 + 17	8 + 18	9 + 19	
0.5				1					1
1.0	1		2	3	1	2	1		10
1.5	4	2	3	5	3	5	8	2	32
2.0	3	4	9	11	15	8	13	4	67
2.5	35	13	17	41	35	13	15	22	191
3.0	43	45	57	41	36	49	39	46	356
3.5	53	44	45	46	47	65	46	60	406
4.0	49	32	55	41	55	47	56	42	377
4.5	31	29	43	33	30	32	40	38	276
5.0	22	19	11	20	18	23	20	28	161
5.5	9	20	7	8	7	6	9	12	78
6.0		17	1	2	3	2	4		29
6.5		12		1	1	2			16
7.0		6					1		7
7.5		4							4
8.0		3							3
8.5		2							2
9.0		1							1
ΣN	250	253	250	253	251	254	252	254	2,017
$\bar{d}$, cm	3.6	4.3	3.6	3.5	3.6	3.7	3.7	3.8	3.7
s_d, cm	0.89	1.43	0.87	1.03	0.97	0.93	1.00	0.87	0.90
v_c %	24.5	32.9	23.7	29.0	26.8	25.1	26.9	23.0	24.3

curves. Figure 116 shows the normal tree distribution (according to Laplace–Gauss), by diameter breast height of *Pinus patula* for various thinning variants at the stand age of 13.5 years.

The difference between the theoretical and empirical distribution of tree frequency was tested using Kolmogorov–Smirnov's criterion according to the formula:

$$\lambda = \frac{d_{max}}{N} .$$

According to this testing criterion, a phenomenon is considered to be fortuitous and not significant, if the computed λ value does not attain the required value $\lambda = 1.36$ for a level of 0.95 reliability.

The computed values for all stand population at a stand age of 2.5–13.5 years did not reach the required value. An only exception was the stand population of the

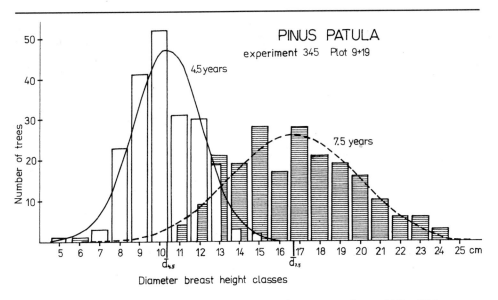

Fig. 115. Empirical and theoretical distribution of *Pinus patula* trees at stand ages of 4.5 and 7.5 years. Thinning plots 9 + 19, experiment 345, Kiwira Forest Reserve, Tanzania.

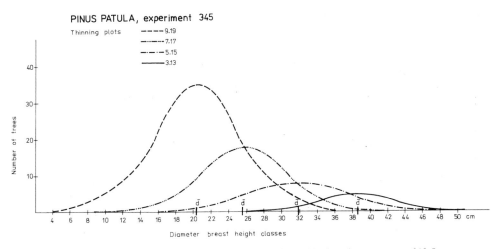

Fig. 116. Normal distribution of tree numbers for *Pinus patula* on thinning plots at an age of 13.5 years. Experiment 345, Kiwira Forest Reserve, Tanzania.

thinning plot 9 + 19 at the stand age of 8.5 years, where the computed value $\lambda = 1.39$ is behind the limit of the required value for the 95% probability.

Number of trees

At the first inventory before the start of any thinning, at a stand age of 2.5 years, there were 1,667–1,693 trees per hectare on the various pairs of thinning plots. At

233

Fig. 117. Stand after very heavy thinning a stand age of 7 years. *Pinus patula,* experiment 345, Kiwira Forest Reserve, Tanzania.

the first thinning, 18% of trees from the thinning plot 9 + 19, and 41–42% of trees from the other plots were removed by cutting.

According to the assigned scheme, at the second thinning 27–30% of the trees were removed and in the next year 28–29% of the trees were taken out, one year later a further 29–30% were cut. Relatively less intense thinnings were carried out at a stand age of 6 years when 21–23% of the trees were removed. In the next year, 35% of the trees were cut on the thinning plot 2 + 12 and 37% of the trees on the plot 3 + 13 were removed. The last and in sequence the seventh thinning, was carried out on the plot 2 + 12 at a stand age of 8.5 years. During this thinning 23% of the trees were removed from the stand.

At the last inventory of thinning plots in Kiwira at a stand age of 13.5 years, the following number of trees remained from the original number: on plot 2 + 12 only 7% of trees, on plot 3 + 13: 10%, on plot 4 + 14: 15%, on plot 5 + 15: 20%, on plot 6 + 16: 25%, on plot 7 + 17: 40%, on plot 8 + 18: 58% and on plot 9 + 19: 81% of trees.

It may be added that very strong thinnings in very young stands were made on all plots with the exception of the pairs of plots 9 + 19 and 8 + 18 (Fig. 117).

234

Mean stand height

The mean stand height was approximately the same on all thinning plots at a stand age of 7.5 years. Beginning from this age, the mean stand height on the pairs of plots 8 + 18 and 9 + 19 with smaller thinning intensities was lower than on plots which had a greater intensity of thinning. If the mean stand height on the plot which had the least thinning is taken as equal to 100% (i.e. plot 9 + 19 at an age of 13.5 years where $\bar{h}$ = 19.5 m), then the mean height on thinning plot 3 + 13 is 113%, on plot 6 + 16 it is 112%, on plot 5 + 15 it equals 111% and on plot 2 + 12 it reaches 110%.

The greater mean height of the stand on plots where there is intensive thinning is apparently the consequence of strong thinnings which in the first place removed trees of lower height and less good quality.

The largest current annual height increment of trees $\Delta \bar{h}$ equals 2.0–2.1 m occurred at the stand age of 3–6 years. During the entire recorded period of 12 years, the current annual height increment did not decline below the value $\Delta \bar{h}$ = 1.1 m.

Mean diameter breast height

The mean diameter breast height of the stand regularly increases with increasing age on all thinning plots. Until a stand age of 4.5 years, the mean diameter breast height of the stand on all thinning plots was approximately equal (d_B* equals 10.5–11.4 cm). Beginning from the stand age of 5.5 years, the values for the mean diameter breast height were lowest on plots with minimum thinning (i.e. 9 + 19), and highest on plots with intensive thinning (i.e. 2 + 12 and 3 + 13) (Fig. 118).

At the last inventory at a stand age of 13.5 years, the following values for the mean diameters breast height were computed on the various thinning plots:

plot 9 + 19	d = 20.9 cm, i.e. 100%,
plot 8 + 18	d = 24.0 cm, i.e. 115%,
plot 7 + 17	d = 25.9 cm, i.e. 124%,
plot 6 + 16	d = 29.2 cm, i.e. 140%,
plot 5 + 15	d = 32.4 cm, i.e. 155%,
plot 4 + 14	d = 34.1 cm, i.e. 163%,
plot 3 + 13	d = 38.8 cm, i.e. 186%,
plot 2 + 12	d = 39.1 cm, i.e. 187%.

Thinning intensity regularly influenced the mean diameter of the stand on the thinning plots.

The current annual diameter increment of the stand on the plot with minimum thinning (i.e. 9 + 19), varied in the range of Δd equal to 0.3–3.3 cm; on the plot with maximum thinning (i.e. 2 + 12), it ranged from 1.7 to 4.9 cm.

---

* d_B has been computed from the mean basal area of the stand.

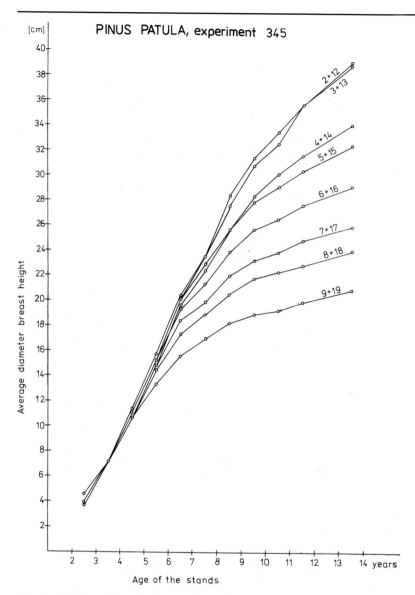

Fig. 118. Position of the mean diameter breast height of the stand of *Pinus patula* as the function of age on thinning plots. Experiment 345, Kiwira Forest Reserve, Tanzania.

The maxima of the current annual diameter increment occurred at a stand age of 6.5 years on plots 2 + 12 to 5 + 15 inclusive (Δd equal to 4.7–4.9 cm).

At 13.5 years, the mean annual diameter increment was 2.9 cm on the most heavily thinned plots (2 + 12 and 3 + 13), 2.4 cm on the medium to heavily thinned plot (5 + 15), 1.9 cm on plot (7 + 17), and 1.5 cm on the least thinned plot (9 + 19). This is an unusually high increment for conditions of the temperate zone.

236

Basal area of the stand

The basal area of the stand, computed in square metres for an area of 1 ha, increased regularly with age only on plots with a lesser thinning intensity (plots 9 + 19, 8 + 18 and 7 + 17).

The large number of trees removed by thinning on plot 2 + 12 caused only a slight increase of the basal area of the stand. Although heavy thinning created an empty space for greater increments in diameter, the small number of trees per unit area did not provide the right conditions for increasing the basal area – which is the main factor promoting timber production in the forest stand.

At the last inventory at an age of 13.5 years, the following values for the basal area of the stand (in m^2 per hectare) were recorded on the various thinning plots:

Basal area	m^2	%	Basal area	m^2	%
plot 9 + 19	47.4	100	plot 5 + 15	28.5	60
plot 8 + 18	43.9	93	plot 4 + 14	23.1	49
plot 7 + 17	36.4	77	plot 3 + 13	20.6	43
plot 6 + 16	33.6	71	plot 2 + 12	14.5	31

If, according to Assmann (1954), on the thinning plots in Kiwira the basal area on plot 9 + 19 at an age of 13.5 years is taken as the maximum, then too heavy a thinning was carried out on plots 6 + 16 to 2 + 12, inclusively. There, the basal area of the stand is smaller than the critical basal area which is required according to Assmann's theory.

Growing stock of the main stand

The growing stock on the thinning plots, i.e. the volume of the main stand was computed by the formula:

$$V = B H F,$$

where V is the growing stock of the stand in m^3, B the basal area of the stand in m^2, H the mean height of the stand in m, F the breast height form factor.

In accordance with the empirical material and for the sake of simplicity of calculation, the value of the form factor was taken as 0.42 for all thinning stands.

At 13.5 years of age the volume of the main stand in m^3 per hectare was as follows for the various thinning plots:

Volume	m^3	%	Volume	m^3	%
plot 9 + 19	388	100	plot 5 + 15	258	66
plot 8 + 18	367	95	plot 4 + 14	203	52
plot 7 + 17	321	83	plot 3 + 13	191	49
plot 6 + 16	309	80	plot 2 + 12	121	34

237

Before starting thinning research, the volume of the main stand was approximately the same on all thinning plots at an age of 2.5 years. After each thinning was made, the number of trees carrying the volume increment decreased and so did the stand production in the wake of a thinning. Where there was no more thinning, timber production continued with increasing age without apparent decline.

Volume of the mean stem of the stand

The volume of the mean stem of the stand was approximately equal on all thinning plots at the start of thinning research.

With increasing age, but mainly as a consequence of thinning process, individual trees with sufficient clear space around then produced much more timber on the heavily thinned plots than did the trees on less intensely thinned plots.

At the last inventory at an age of 13.5 years, the volume of the mean stem on the various thinning plots (in m^3) was as follows:

Volume	m^3	%	Volume	m^3	%
plot 9 + 19	0.281	100	plot 5 + 15	0.744	265
plot 8 + 18	0.377	134	plot 4 + 14	0.802	285
plot 7 + 17	0.463	165	plot 3 + 13	1.104	393
plot 6 + 16	0.627	223	plot 2 + 12	1.092	389

The volume of the mean stem on plots with very heavy thinning was much higher at the last stand inventory than it was on plots with a less intensive thinning.

Total production

The total production of the stand (i.e. the volume of the main stand at a given age and the sum total of all intermediate cuttings), is of the greatest importance in judging the influence of thinnings on the growth potential of stands.

Numerical data on the total production of all thinning plots at a stand age of 13.5 years, i.e. after 11 inventories were made, are given in Table 59.

The largest total timber production occurred on thinning plot 9 + 19, i.e. on the plot with the smallest thinning intensity. On all other thinning plots the remaining trees in the stand produced less timber. Compared to the thinning plot 9 + 19, the losses of total timber production were found to be 5% on plot 8 + 18 with one heavy thinning, 16% on plot 7 + 17 with two heavy thinnings, 18% on plot 6 + 16 with three heavy thinnings, 29% on plot 5 + 15 with four heavy thinnings, 41% on plot 4 + 14 with five heavy thinnings, 38% on plot 3 + 13 with six heavy thinnings, and a total production loss of 51% on plot 2 + 12 where seven heavy thinnings were carried out during a period of seven years.

238

Table 59. *Pinus patula,* Experiment 345. Total volume production per hectare of stand, inventoried at an age of 13.5 years

Thinned plots	Volume of main stand	Cumulative volume from thinnings	Total timber production	
number	m³ per ha	m³ per ha	m³ per ha	%
9 + 19	388	–	388	100
8 + 18	367	1	368	95
7 + 17	321	3	324	84
6 + 16	309	9	318	82
5 + 15	258	18	276	71
4 + 14	203	26	229	59
3 + 13	191	51	242	62
2 + 12	131	61	192	49

Total current annual volume increment

The total current annual volume increment reached its maximum at a stand age of 9 years on almost all thinning plots. The largest values of the total current increment occurred on plots with a lighter thinning, namely 50 m³ on plot 9 + 19, 46 m³ on plot 8 + 18, and 42 m³ per hectare on plot 7 + 17. In contrast, on plot 2 + 12 with the heaviest thinning, at the same age, the value of the total current increment was 19 m³ per hectare.

From the age of 10 years onwards the total current annual volume increment declined gradually on all thinning plots.

Thus numerical data concerning total volume production and total current increment on the thinning plots draw attention to the losses which occur in the total volume production of timber on plots with very heavy thinning regimes.

6.2 Thinning experimental plots in Ngwazi

6.2.1 Methodology

The stand is situated in the Sao Hill in a mountain savanna at an altitude of 1,700 m. The average annual rainfall is 1,300 mm and the rainy season lasts from December to May. The soil is loamy and greyish-brown. The stand was established in January 1960 using strong plants in a spacing of 2.7 by 2.4 m (9 feet by 8 feet). The experimental plots were demarcated in August 1964, i.e. at a stand age of 4.5 years.

In the artificially established stand age of 4.5 years, the area intended for thinning research was divided into four blocks – I, II, III and IV. In each block four thinning plots A, B, C and D with different thinning intensities were chosen at random. The

239

area of each plot was 0.13 ha (0.31–0.32 acres), which results in an aggregate of 0.52 ha for each group of four plots, i.e. I-A to IV-A; I-B to IV-B; I-C to IV-C and 1-D to IV-D. The total area of the plots was thus 2.08 ha. The plots border on isolation belts (Fig. 119). There are four thinning variants, namely A with no thinning, B with one, C with three and D with five thinnings. The prescribed thinnings are apparent from Table 60.

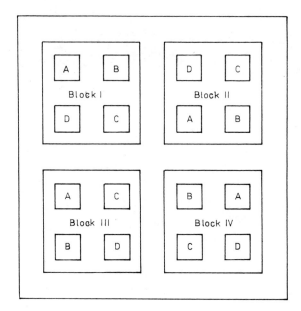

Fig. 119. Pattern of the distribution of thinning plots for *Pinus patula*. Experiment 661, Sao Hill, Tanzania.

Table 60. Scheme of thinning variants A–D at Ngwazi, Experiment 661

Thinning variants		Number of trees after thinning in a year					
	per ha	1,540	770	500	390	250	190
	per acre	600	300	200	150	100	75
	%	100	50	32.4	26.3	16.2	12.3
A	1960	not thinned					
B	1960	1964	–	–	–	–	
C	1960	1964	1965	1966	–	–	
D	1960	1964	1965	1966	1967	1968	

1 acre = approximately 0.404 ha.

Thinning research was based on having a constant number of trees in the stand at the end of the prescribed number of thinnings. On plots A no thinning was done (Fig. 120); these are control plots where the number of trees at the mature cutting age should be approximately the same as it was at the time of establishment (i.e. 192 trees on the plot, or 1,540 trees per hectare). On plots B after all thinnings have

240

Fig. 120. *Pinus patula,* experiment 661, age: 10 years. Not thinned plot. Ngwazi region of Sao Hill, Tanzania.

been made the number of trees should be 96 (i.e. 770 per hectare), on plots C it should be 48 trees (i.e. 390 trees per hectare), and on plots D a total of 24 trees (i.e. 190 trees per hectare) should remain.

The first pruning of the stand to one half of the tree height was made at a stand age of 3.5 years. At that time the trees had a height of 5–6 m.

In marking the thinnings a rule was applied that trees of good quality in the stand should be retained and that trees of inferior quality should be removed making due allowance for the uniform distribution of trees in the stand. All trees on the thinning plots were numbered and the height of 1.3 m above the ground was marked on them. Diameters were measured for all trees using a steel tape measure adapted to this purpose. The heights of randomly selected trees were measured using a Blume–Leiss hypsometer and included at least 10% of the trees on the thinning plot.

The thinning plots before and after the thinning operations and also the thinnings themselves were carefully measured and the data recorded. This allowed for their cruising evaluation to be done by using statistical mathematical methods.

6.2.2 Analysis of results

An analysis of the dispersion of diameters breast height of trees between the various thinning plots was made on the basis of numerical data at the stand age of 4.5 and 13.5 years. For testing for the 95% level of significance Fischer–Snedecor's distribution (i.e. the F test), was used (Myslivec, 1957).

Whereas the differences between the thinning plots at a stand age of 4.5 years were statistically not significant, the analysis of numerical values at a stand age of 13.5 years showed that unsignificant difference only occurred between all thinning plots B and C.

Significant differences existed on plots with the heavy thinning treatment D, namely between the plots I-D ~ II-D, II-D ~ III-D and II-D ~ IV-D. They were caused by the thinning plot II-D – the numerical data of which for dispersion, standard deviation and variation coefficient are lower, and the variation width of the stand population is smaller than the numerical data of plots I-D, III-D and IV-D.

By analogy also, plots I-A and II-A (which are of poorer growth than plots III-A and IV-A), may be considered as control plots. The computed F values only slightly exceed the critical values of $F_{0.05}$.

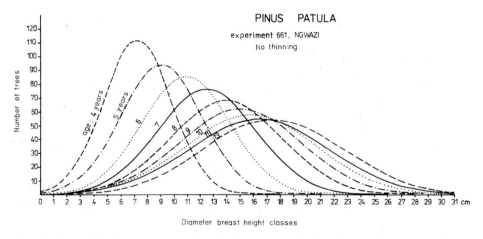

Fig. 121. Normal distribution of tree numbers by diameter breast height. *Pinus patula*, plots A – without thinning. Experiment 661, Ngwazi, Tanzania.

In order to express the diameter distribution of tree frequencies in the investigated stand populations and in the variously marked thinnings, as well as their influence on the diameter distribution of the stand population, then the normal Laplace–Gauss distribution of tree frequencies was used as a mathematical model. The computed statistical characteristics of the various populations were compared to each other.

The normal distribution of tree frequencies in the various years of measurement has a regular course, whether this refers to the stand populations before or after thinning which are represented above the axis x, or whether thinning populations are considered as is shown by frequency curves below the axis x (Figs 121, 122).

The main statistical characteristics of stand populations before thinning at the ages of 4.5, 8.5, 10.5 and 13.5 years are given in Table 61.

Table 61. *Pinus patula*. Experiment 661. The main statistical characteristics of stand population on the thinned areas

Thinned plots	Number of trees	Mean diameter $\bar{d}$	Standard deviation s_d	Coefficient of variability v_c
		cm	cm	%
	age: 4.5 years			
A	711	7.3	2.53	44.7
B	722	7.1	2.41	34.1
C	731	7.5	2.26	30.1
D	740	7.3	2.30	31.5
	age: 8.5 years			
A	710	13.9	4.15	29.8
B	386	16.4	3.49	21.3
C	198	18.4	2.83	15.3
D	130	19.6	3.17	16.3
	age: 10.5 years			
A	710	15.8	4.88	30.9
B	386	19.5	4.16	21.4
C	198	22.4	3.51	15.7
D	96	25.9	4.36	16.8
	age: 13.5 years			
A	694	17.6	5.23	29.7
B	385	21.9	4.86	22.2
C	197	25.9	4.20	16.2
D	95	31.4	5.07	16.2

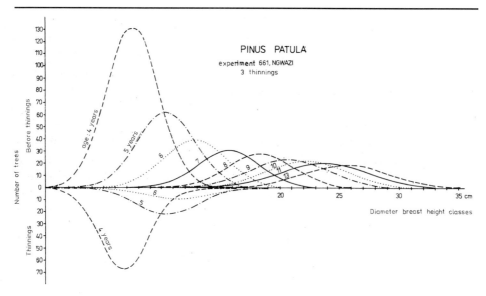

Fig. 122. Normal distribution of tree numbers by diameter breast height. *Pinus patula,* plots C – with three thinnings. Experiment 661, Ngwazi, Tanzania.

Number of trees

At the time of establishing the thinning research (at a stand age of 4.5 years), the actual number of trees on the various thinning plots did not correspond to the prescribed initial number of 1,540 trees per hectare (192 trees on each plot). It was as follows: the equivalent 1,380 trees per hectare on plot A, 1,399 trees on plot B, 1,432 trees on plot C, and 1,434 trees on plot D.

On the four control thinning plots A (without thinning) there was no natural loss between the ages of 4.5–7.5 years. Up to the stand age of 11.5 years, i.e. during the seven year period of recording, four overtopped trees perished, i.e. 8 trees per hectare. In the next two years the natural loss of trees was much higher (32 trees per hectare). This means that the natural loss of trees in the stand without thinning between the ages of 4.5–13.5 years amounted to 3%.

On plot B only one thinning was undertaken, this being done at a stand age of 4.5 years. In this thinning the equivalent of 651 trees per hectare were cut, i.e. 47% of the total number of trees on the thinning plot. As to intensity, this thinning must be considered to be a heavy one.

On plot C three thinnings in succession were made at stand ages of 4.5–6.5 years. In the first thinning the equivalent of 688 trees per hectare were cut (i.e. 48% of all trees), in the second thinning 234 trees per hectare (i.e. 31% of all trees), and in the third thinning 124 trees per hectare (i.e. 24% of all trees) on plot C were removed by felling. All three thinnings should be classified as heavy ones. From the original number of 1,432 trees per hectare there remained after three thinnings the equivalent of 384 trees per hectare (i.e. 27% of the original number of trees).

244

On plot D a total of five thinnings were carried out between stand ages of 4.5–8.5 years. At the first thinning intervention the equivalent of 692 trees per hectare (i.e. 48% of all trees) were removed. The following removals being: 244 trees per hectare (i.e. 33% of trees) at the second thinning, 126 trees per hectare (i.e. 25% of trees) at the third thinning, 120 trees per hectare (i.e. 32% of trees) at the fourth thinning, and 66 trees per hectare (i.e. 26% of trees) at the fifth thinning. After five thinnings the equivalent of 186 trees per hectare (i.e. 13% of the original number of trees) remained in the stand at the age of 8.5 years.

Basically, the thinning intensity was approximately uniform with an annual thinning interval but with a different number of thinnings.

Mean height of the stand

The mean height of the stand was approximately the same on all thinning plots both before beginning the thinning at a stand age 4.5 years, and after ten years' measurement at a stand age of 13.5 years. At the age of 4.5 years the mean height of the stand on the various thinning plots was 5.6–5.8 m, and at the age of 13.5 years it was 17.4–18.0 m.

It may be claimed that neither the intensity nor the number of thinnings had any important effect on the mean height of the stand or on its site class.

During the observational period of ten years the average annual height increment did not decrease below the values of 1.2 m on any plot (Table 62).

Table 62. *Pinus patula,* Experiment 661. Data of mean annual height increments in thinning plots between the ages of 4.5–13.5 years

Age of stand	Mean annual height increment in metres			
	Thinning plots			
years	A	B	C	D
4.5	1.29	1.27	1.27	1.24
5.5	1.25	1.22	1.20	1.20
6.5	1.29	1.23	1.22	1.28
7.5	1.47	1.44	1.43	1.39
8.5	1.41	1.34	1.33	1.41
9.5	1.43	1.43	1.42	1.39
10.5	1.45	1.44	1.41	1.36
11.5	1.41	1.37	1.36	1.37
13.5	1.32	1.37	1.29	1.29

Mean diameter of the stand

The mean diameter breast height of the stand computed from the mean basal area of the stand as the function of age are represented graphically in Fig. 123.

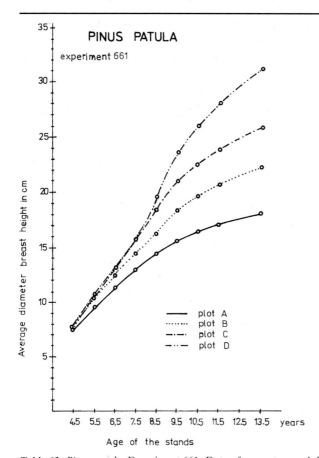

PINUS PATULA

experiment 661

Average diameter breast height in cm

plot A
plot B
plot C
plot D

4,5 5,5 6,5 7,5 8,5 9,5 10,5 11,5 12,5 13,5   years

Age of the stands

Fig. 123. Position of the mean diameter breast height of the stand as the function of age on thinning plots of *Pinus patula*. Experiment 661, Ngwazi, Tanzania.

Table 63. *Pinus patula,* Experiment 661. Data of current annual diameter increments in thinning plots between the ages of 5.5–13.5 years

Age of stand	Current annual diameter increment in cm			
	Thinning plots			
years	A	B	C	D
5.5	1.9	2.4	2.2	2.3
6.5	1.8	2.1	2.3	2.3
7.5	1.6	2.0	2.4	2.3
8.5	1.5	1.9	2.7	3.1
9.5	1.2	2.2	2.6	3.5
10.5	0.8	1.3	1.5	2.4
11.5	0.7	1.1	1.4	2.1
13.5	0.6	0.8	1.1	1.6

The values of the mean diameter breast height of the stand are the lowest in the stand without thinning; they gradually increase with the diminishing number of trees on the thinning plots. The largest values of mean diameter breast height are on plots

246

D with five thinnings which reduced the original number of trees in the stand to 13%.

Whereas at a stand age of 4.5 years (i.e. before the first thinnings), the mean diameter breast height of the stand was almost equal on all thinning plots (d equals 7.5–7.8 cm), at the age of 13.5 years (i.e. at the last inventory), the values of the mean diameter on the various thinning plots were as follows: plot A 18.4 cm (i.e. 100%), plot B 22.5 cm (i.e. 122%), plot C 26.2 cm (i.e. 142%), and plot D 31.6 cm (i.e. 172%).

The direct influence of thinning (i.e. of the reduction in tree numbers) on the increase of the diameter increment is evident on all three thinning plots (Table 63).

Basal area of the stand

The basal area of the stand increased with greater age viz. regularly on the control plot A and on plot B with one thinning. On plots C and D with heavy thinning the basal area of the stand at a younger age did not increase because of the removal of an unproportionally large number of trees from the stand.

The highest values of the basal area of the stand were obtained at the last inventory at a stand age of 13.5 years. On the control plot A the basal area was 35.6 m^2 per hectare (= 100%). The comparison of the basal area data of plot A with the basal area data from the other three thinning plots shows that at an age of 13.5 years the basal area on plot B was 29.5 m^2 per hectare (i.e. 83% of the basal area of stand A without thinning); on plot C it was 20.6 m^2 per hectare (i.e. 58%); and on plot D it was 14.7 m^2 per hectare (i.e. only 41%).

If the hypothesis is accepted that the basal area on thinning plot B at an age of 13.5 years corresponds to the optimum basal area of the stand, then according to Assmann's theory (Assmann, 1954) the values of the basal area on plot C (58%) and on plot D (41%) are demonstrably below the critical basal area. The critical basal area corresponds, according to Assmann, to about 75% of the maximum basal area of the stand which may be identified as that on plot A.

Volume of the main stand

At a stand age of 13.5 years the volume of the main stand in m^3 per hectare on the various thinning plots was as follows:

Volume	m^3	%
thinning plot A	266	100
thinning plot B	223	84
thinning plot C	151	57
thinning plot D	108	41

The relative estimation of the timber production of the main stand shows that the percentage shares are similar to those for the basal area on the various thinning plots. The largest amount of timber for an equal time period was produced on plot A without thinning and the smallest amount on plot D where the number of trees was reduced to 13% of the original number as a consequence of five thinning operations.

Volume of the mean stem of the stand

Before the start of thinning research at a stand age of 4.5 years the volume of the mean stem (V) of the stand was approximately the same for all four thinning blocks, viz. $V \doteq 0.009 \text{ m}^3$.

At the age of 13.5 years, after the last inventory, the volume of the mean stem of the stand in m^3 on the thinning plots was as follows:

Volume	m^3	%
thinning plot A	0.198	100
thinning plot B	0.300	152
thinning plot C	0.397	200
thinning plot D	0.590	298

The much higher volume of the mean stem on plots with intensive thinning is a consequence of increased increment of released trees in the stand, i.e. the consequence of heavier thinning. When opening up the crown cover and admitting light, mainly on plot D, the disengaged trees being set free produce much more timber than do trees in a fully stocked stand with no thinning.

Total timber production

The total timber production on the various thinning plots at the age of 13.5 years is given in Table 64. As can be seen from this table, the largest total production

Table 64. *Pinus patula*, Experiment 661. Total volume production in the thinned stands aged 13.5 years

Thinned plots	Volume of main stand	Cumulative volume from thinnings	Total timber production	
	m^3 per ha		m^3 per ha	%
A	266	–	266	100
B	223	5	228	86
C	151	15	166	62
D	108	32	140	53

during the period observed occurs on plot A without thinning. As compared to this control plot, the total losses of total timber production at this age amount to 14% on plot B with one thinning, to 38% on plot C with three thinnings, and to 47% on plot D with five thinnings.

Total current annual volume increment

The total current annual volume increment attains its maximum at a stand age of 9.5 years on all four thinning plots.

At this age the total current annual increment was 45 m^3 per hectare on control plot A, 38 m^3 per hectare on plot B, 33 m^3 per hectare on plot C, and 32 m^3 per hectare on thinning plot D. Afterwards, from the age of 9.5 years on, the total current annual volume increment gradually declines on all thinning plots.

Important survey indices (such as the number of trees on the unit area, the basal area and the total current volume increment), reveal the fact that very heavy thinning made at a young age of the investigated artificially established stand of *Pinus patula* caused a reduction of the total timber production on all three plots with thinnings, as compared to the control plot without thinning.

6.3 Comparison of thinning research in *Pinus patula* in Kiwira and Ngwazi

The stands of *Pinus patula* in Kiwira experimental plot No. 345 and in Ngwazi experimental plot No. 661, where the influence of thinnings of different intensity on the growth potential of the stand has been investigated, were both established in January 1960. They are therefore of equal age, thus enabling the mutual comparison of survey data.

To match the four thinning variants A, B, C and D in the stand at Ngwazi (experiment 661), the author has selected from the stand at Kiwira (experiment 345), such pairs of thinning plots which (according to the number of trees remaining per hectare after the last thinning) approximately corresponded to the plots in Ngwazi. They are thinning plots 9 + 19,7 + 17,6 + 16 and 3 + 13.

The main survey data for the various pairs of thinning plots at stand ages of 4.5, 9.5 and 13.5 years are given in Table 65.

Comparing the survey data from both pairs of investigated stands it is apparent that the artificially established stand of *Pinus patula* in Kiwira (experiment 345), was more developed as to height, diameter and volume than was the pine stand in Ngwazi (experiment 661).

The thinning plots in Kiwira and Ngwazi were established on the basis that there is a constant number of trees in the stand from the last thinning to the end of the rotation, including the control plots without thinning intervention. It was assumed that the released trees of *Pinus patula* would produce more timber than would those trees growing in a denser canopy (O'Connor, 1935).

Fig. 124. Forest nursery near plantations of *Pinus patula*. Sao Hill, Tanzania.

In spite of the fact that the equally aged stands of *Pinus patula* investigated in Kiwira and Ngwazi differ as to growth and that the stand in Kiwira is more developed in diameter, height and timber volume; common conclusions can be made for both stands with regard to the analogousness of thinnings.

For the expression of the diameter distribution of trees on the thinning plots the author used suitable basic statistical characteristics, namely the arithmetical mean and the standard deviation. By means of them he determined the normal frequency distribution of trees. From a stand age of 8.5 years, the differences between the empirical and the theoretical distribution were not significant on all thinning plots in Ngwazi (experiment 661). In Kiwira, a stand of a better site class, these differences according to Kolmogorov–Smirnov's criterion were not significant on all thinning plots in all inventories (Borota, 1980).

The mean height of the stand was not influenced substantially either by the intensity of thinning, or by the larger number of thinnings on the thinning plots in Ngwazi; but in the thinning stand at Kiwira it seems that the greater height (about 1.5–2.5 m – at 13.5 years of age), on a very opened up plot is the consequence of heavy thinning.

The mean diameter breast height has a regular course. The lowest values were on plots with the smallest number of thinnings, the highest ones on plots with the largest number of heavy thinnings which reduced the original number of trees to a minimum.

The basal area of the stand, as the main timber-creating factor, regularly increased

Table 65. A comparison of enumerated data in the thinned stands of *Pinus patula* at Ngwazi, experiment 661 and at Kiwira, experiment 345

Data per hectare	Thinned plots							
	A	9 + 19	B	7 + 17	C	5 + 15	D	3 + 13
Age of the stand: 4.5 years								
Number of trees	1,380	1,387	1,399	693	1432	693	1434	693
Main height, m	5.8	6.6	5.7	7.1	5.7	7.3	5.6	7.4
Main D.B.H., cm	7.7	10.5	7.5	11.1	7.8	11.4	7.7	11.4
Basal area, m²	6.5	12.1	6.1	6.8	6.9	7.1	6.5	7.1
Volume, m³	13	33	13	20	14	22	13	22
Volume of main stem, m³	0.009	0.024	0.009	0.029	0.009	0.032	0.009	0.032
Age of the stand: 9.5 years								
Number of trees	1,376	1,380	748	693	384	347	186	173
Main height, m	13.6	15.9	13.6	16.5	13.5	16.9	13.2	17.2
Main D.B.H., cm	15.7	18.9	18.6	23.2	21.2	27.9	23.9	31.4
Basal area, m²	26.6	39.0	20.3	29.0	13.4	21.2	8.4	13.5
Volume, m³	163	261	123	201	83	150	54	97
Volume of main stem, m³	0.118	0.189	0.164	0.290	0.216	0.432	0.290	0.561
Age of the stand: 13.5 years								
Number of trees	1,340	1,380	743	693	380	347	183	173
Main height, m	17.8	19.5	18.0	21.0	17.4	21.6	17.4	22.1
Main D.B.H., cm	18.4	20.9	22.5	25.9	26.2	32.4	31.6	38.8
Basal area, m²	35.6	47.4	29.5	36.4	20.6	28.5	14.7	20.6
Volume, m³	266	388	223	321	151	258	108	191
Volume of main stem, m³	0.198	0.281	0.300	0.463	0.397	0.744	0.590	1.104
Total volume, m³	266	388	228	324	166	276	140	242

with the increasing age of the stand on thinning plots with a smaller intensity of thinning. On plots with intensive thinnings the basal area of the stand increased only slightly until the end of thinnings, as a consequence of the large number of trees which were being removed from the plots.

The volume of the mean tree of the stand was approximately equal at the beginning of the thinning experimental on all thinning plots in both stands.

In considering the influence of thinnings on the growth potential of forest stands then the total timber production (i.e. the volume of the main stand at a given age and the sum total of all thinnings) is of decisive importance.

Data on the total timber production obtained from the last inventory at a stand age of 13.5 years is listed in Tables 59 and 64. This shows that plots without thinning (plot A – experiment 661) and plots with the smallest thinning intensity (plots 9 + 19 and 8 + 18 – experiment 345) were the most productive. Production was lowest on thinning plots where the largest percentage of trees was removed from the stand by cutting.

The survey data listed underline the fact that very heavy thinnings made at a young age in artificially established stands of *Pinus patula* caused a decrease in the total timber production.

With regard to the tree dimensions obtained on control plots, or on plots with slight thinning, it appears that it is not reasonable to make the first thinning before a stand age of 8–10 years, and then about 30% of the trees should be removed from the stand. The next three to four equally heavy thinnings should be made at five-year intervals. It may be assumed that the material obtained from intermediate cuttings could be sold on the local market.

7 Conclusions

Tropical forests as the most productive biomes of the Earth have got into the limelight of world attention both from the ecological and the timber production viewpoints.

Annually at present more than 3 billion m^3 of timber is cut and consumed world-wide. This volume includes about 46% of industrial wood and 54% of fuelwood. About 48% of the world timber production belongs to the developing countries. However, developing countries contribute only about 17% (1985) of the total world production of industrial wood. In addition, the constant growth of the human population in the developing countries also increases their share of the consumption of wood, mostly for fuel, and also the area of forest available for each inhabitant is decreased. At present, developing countries participate with about 73% in the world's production of fuelwood.

Developing countries have an important share in the export of logs of broad-leaved species (89%) and broad-leaved sawn wood (68% of world export in 1985). Also, the share of developing countries in the world export of veneer sheets (53%) and plywood (70%) is high. This testifies to the well-established plywood industry in some developing countries (Brazil, Indonesia, South Korea).

A substantial part of this book has been concerned with the analysis of species composition and the diameter structure of trees in natural tropical forests in selected localities in Ghana, Congo, Gabon, Laos and Tanzania. In these countries it is important to obtain information on the evaluation of basic data about the raw materials existing in tropical forests, on their tree species composition, on the assortment structure of trees and their production potential. This information is very useful for forestry personnel working in developing countries, and it acts as an impulse and an asset for scientific research in the tropics.

The analysis of empirical material confirmed that differing ecological requirements of tree species in natural tropical forests influence the multiform growth of trees, their characteristical diameter and height structure, the space distribution of trees and the occurrence of many tree species in the forest stand. Some tree species grow in groups, others occur in the forest sporadically and belong only to certain forest communities. The number of trees of various species on a given unit area in the natural tropical forest is, in distinct localities, very different. Also the volume of trees mature enough for cutting in the same tree species is variable. The most valuable commercial tree species suitable, for example, for the production of decorative veneer sheets occur mainly in moist semi-deciduous tropical forests and not in the evergreen moist tropical forests as has been hitherto generally believed.

The materials investigated have proved that the mathematical formula of the beta function used in this work is suitable for the theoretical expression of tree distribution by diameter breast height and tree height in natural tropical forests. In most cases a decreasing tendency of tree distribution for the various tree species, or timber categories, from the smallest to the largest trees was observed. Also in the dry deciduous tropical forests of the "miombo" type in Tanzania (on the localities studied), the tree distribution by diameter of whole stands corresponded to the distribution according to the beta function. The frequency curves revealed a decreasing tendency with an increasing diameter breast height of the trees. But the diameter distribution of trees for certain commercial tree species in these stands corresponded approximately to the normal frequency distribution of trees according to Laplace–Gauss. The occurrence of a small number of trees in smaller diameter classes is probably the consequence of forest fires which almost regularly each year cause damage to the ecosystem, destroy the microflora and inflict damage to young trees thus frequently preventing the regeneration of forest stands.

The declining tendency of tree distribution for the species *Pterocarpus angolensis* from the smallest values of the annual diameter increment to the largest ones is unprecedented, and differs entirely from tree distribution according to their diameter increment in other forest stands which follow the normal Laplace–Gauss distribution. In the dry deciduous tropical forests of the "miombo" type tree growth is slow. The current annual diameter increment of trees in the "miombo" forests is smaller than the diameter increment of tree species growing in forests of the temperate zone on average site classes.

Special attention has been paid in this book to the growth processes in forest plantations of certain fast growing tree species which are cultivated in the tropics and subtropics for fuelwood or for timber used in the wood-processing industries. In the book an evaluation is given of the dendrometrical measurements which have been recorded on permanent experimental plots and the diameter and height structure of trees and stands and their increment are analysed.

The analysis of numerical data obtained from these even-aged stands of fast growing tree species confirmed that the distribution of trees with regard to their diameter breast height, height and increment, approaches the normal Laplace–Gauss distribution. This refers to stands of advanced production with a high annual increment. The analysis of the growth of *Shorea robusta* confirmed that trees mostly conserve their hierarchical position during their life-time.

A separate chapter is devoted to thinnings made in artificially established stands of *Pinus patula* in Tanzania. Numerical data obtained from measurements of trees

254

on permanent thinning plots were analysed and evaluated using statistical mathematical methods. On the basis of this analysis of the main survey values, conclusions have been made and recommendations formulated for the realization of results in the practice of growing tree plantations in developing countries. Numerical data about the total timber production and the total current increment from experimental thinning plots, disclose losses in the total timber production which occurs in stands with very heavy thinning.

The book with all its data and information is intended to serve as a guide for promoting an economically rational and ecologically sound management and use of tropical forests in developing countries.

References

ADDO, S., BAIDOE, J. F. (1962), Working plan for Attandaso Group Forest Reserves, Cape Coast Forest District, Accra.

AGGARWAL, K. L. (1949), Fourth Working Plan for the Kulu and Seraj Forests (Punjab) 1949/50–1979/80, Kulu Forest Division.

AHVENINEN, H. (1979), Export Market Prospects for Laotian Wood and Wood Products, Jaakko Pöyry, Helsinki.

ANDOH, L. I. et al. (1965), Working Plan for Tain Group Forest Reserve, Sunyani Forest District, Accra.

ASSMANN, E. (1954), Grundflächenhaltung und Zuwachsleistung Bayerischer Fichten-Durchforstungsreihen, Forstwiss. Cbl. **93**, 4, 177–179.

ASSMANN, E. (1961), Waldertragskunde, BLV Verlagsgesellschaft, München–Bonn–Wien, 490 pp.

AUBRÉVILLE, A. (1961), Étude écologique des principales formations végétales du Brésil et contribution à la connaissance des forêts de l'Amazonie brésilienne, Centre Technique Forestier Tropical, Nogent sur Marne, 268 pp.

AUBRÉVILLE, A. M. A. (1985), The disappearance of the tropical forests of Africa, Unasylva **37**, 148, 18–27.

BAIDOE, J. F. (1964), Working Plan for Bosumkese Group Forest Reserves, Sunyani Forest District, Accra.

BANERJI, J. (1957), Tropical Rain Forest, in Tropical Silviculture, FAO, Rome, pp. 36–45.

BAUR, G. N. (1964), Rain forest treatment, Unasylva **16**, 72, 18–26.

BEARD, J. S. (1955), The classification of tropical American vegetation types, Ecology **36**, 89–100.

BERHANU, L. (1984), Silviculture and growth-rate of eucalypts in Ethiopia, Dipl. Thesis, University College of Forestry and Wood Technology, Zvolen, 77 pp. (in Slovak).

BERTRAND, A. (1974), Marché des grumes et des sciages africains en Europe, Bois et Forêts des Tropiques 153, 43–60.

BETHEL, J. S. (1984), Sometimes the world is "weed": A critical look at lesser-known species, Unasylva **36**, 145, 17–22.

BODEN, R. W. (1964), Hybridisation in *Eucalyptus,* Indian Forester **90**, 9, 581–582.

BOROTA, J. (1960), Management evaluation of thinnings and increment in spruce stands, in Mathematical-Statistical Methods in Forest Management and Silviculture, Scientific Papers of the Slovak Academy of Sciences, Bratislava, pp. 187–263 (in Slovak).

BOROTA, J. (1965), Some notes on the growth and increment of the hundred-year-old plantation of *Eucalyptus globulus* in the Nilgiris Hills, Indian Forester **91**, 6, 418–422.

BOROTA, J. (1966), Inventory of *Juniperus procera* at Magamba and Shume, Silvic. Section of Forest Division, Lushoto, Tanzania, 26 pp. (unpublished).

BOROTA, J. (1969), The eucalypts at Lushoto Arboretum, Tanzania Silvicult. Research Note 12, 1–11.

256

BOROTA, J. (1971a), Annual reports of the silviculturist 1969–1970, Silvicult. Research Section, Lushoto, 32 pp.

BOROTA, J. (1971b), The results of the oldest thinning experiment in *Pinus patula*, Tanzania Silvicult. Research Note 18, 1–7.

BOROTA, J. (1971c), The growth of tree species in Lushoto Arboretum, Tanzania Silvicult, Research Note 23, 1–24.

BOROTA, J. (1972), The silviculture and growth of *Pinus caribaea* MORELET in Tanzania, Scientific Papers of the Faculty of Wood Technology, University College of Forestry and Wood Technology, Zvolen, pp. 59–69.

BOROTA, J. (1973), Aménagement du Massif Forestier, Congo, Rapport technique No. 4 SF/PRC-15, FAO, Rome, 66 pp.

BOROTA, J. (1980), Influence of thinning on the growth of *Pinus patula* in artificially established stands, Lesnictví **26**, 11, 981–1008 (in Slovak).

BOROTA, J. (1981), Composition and diameter structure of high tropical forests on the example of Ghana, Acta Facultatis Forestalis Vol. 23, Zvolen, pp. 43–62 (in Slovak).

BOROTA, J. (1982), The Himalayan cedar forests, Silvaecultura tropica et subtropica **9**, 113–124.

BOROTA, J. (1984), Composition and structure of tropical forests, Dissertation, University College of Forestry and Wood Technology, Zvolen, 333 pp. (in Slovak).

BOROTA, J. (1987), Rapport sur la Mission en Aménagement dans les U.F.A. 2 et 5, FAO Document de terrain, No. 25, Brazzaville, 44 pp.

BOROTA, J., PROCTER, J. (1967), A review of softwood thinning practice and research in Tanzania, Paper written for FAO Symposium on Man-made Forests, Canberra, Australia, 8 pp.

BOROTA, J., PERSSON, A. (1971), Growing eucalypts for poles at Olmotonyi, Arusha region, Technical Note No. 16, Lushoto, Tanzania, pp. 1–8.

BROWN, A. et al. (1968), Growing Trees on Australian Farms, Commonwealth Government Printer, Canberra, 397 pp.

BRÜNING, E. P. (1970), Stand structure, physiology and environmental factors in some lowland forests in Sarawak, Tropical Ecology **II**, 26–43.

BRYCE, J. M. (1967), The Commercial Timbers of Tanzania, Tanzania Forest Division, Utilization Section, Moshi, 139 pp.

BURGESS, P. F. (1961), The structure and composition of lowland tropical rain forest in northern Borneo, Malayan Forester **24**, 1, 68–80.

BURLEY, J., NIKLES, D. G. (1972, 1973), Selection and Breeding to Improve Some Tropical Conifers, Commonwealth Forestry Institute, Oxford, Vols 1 and 2.

BUTTOUD, G., HAMADOU, M. (1986), Trade in forest products among developing countries, Unasylva **38**, 153, 20–27.

CATINOT, R. (1978), The forest ecosystems of Gabon, in Tropical Forest Ecosystems, UNESCO–UNEP–FAO, Paris, pp. 575–579.

CHAMPION, H. G. (1936), A preliminary survey of the forest types of India, Reprinted in 1961, Indian Forest Records, N. S., Silviculture **1**, 1, 204 pp.

CHAMPION, H. G., GRIFFITH, A. L. (1960), Manual of General Silviculture for India, 3rd ed., Delhi.

CHAUDRY, M. A., SILIM, S. (1980), Agri-silviculture in Uganda, Unasylva **32**, 128, 21–25.

CHRISTENSEN, B. (1983), Mangroves – what are they worth? Unasylva **35**, 139, 2–15.

COUSENS, J. E. (1965), Some reflections on the nature of Malayan lowland rain forest, Malayan Forester **28**, 122–128.

CRAIB, I. J. (1939), Thinning, pruning and management studies on the main exotic conifers grown in South Africa, Union of South Africa, Dept. Agric. Fore. Sci. Bull. No. 196.

ČERNÝ, J. et al. (1980), Report from a business trip to Laos, Brandýs nad Labem, 95 pp. (in Czech, unpublished).

DAVIS, T. A. W., RICHARDS, P. W. (1933, 1934), The vegetation of Moraballi Creek, British Guiana, J. Ecology **21**, 350–384; **22**, 106–155.

DAWKINS H. C. (1958), The management of natural tropical high forest with special reference to Uganda, Institute paper No. 34, Imperial Forestry Institute, Oxford, 155 pp.

DAWSON, G. A. et al. (1972), Ghana Forest Product Transport Study, Canadian International Development Agency, Accra.

DOLEŽAL, B., KORF, V., PRIESOL, A. (1969), Management of Forests, SZN, Prague, 405 pp. (in Czech).

ECE (1974), Study of timber and prospects in the ECE region 1950–2000, Timber Committee ECE, Geneva.

ECE (1978–1986), Timber Bulletin for Europe, ECE, Geneva.

ECE (1981), Annual forest products market review, Timber Bulletin for Europe, Vol. 34, Suppl. 1, Economic Commission for Europe, Geneva.

ERFURTH, T., RUSCHE, H. (1976), The Marketing of Tropical Wood, Wood Species from African Tropical Moist Forests, FAO, Rome.

ERNST, W. (1971), Zur Ökologie der Miombo-Wälder, Flora 160, 317–331.

FAO (1960), World Forest Inventory – 1958, Rome.

FAO (1966), World Forest Inventory – 1963, Rome.

FAO (1970), Développement forestier – Gabon, I rapport, Rome.

FAO (1975–1986), Yearbook of Forest Products, FAO, Rome.

FAO (1981), Forest Resources of Tropical Africa, Parts 1 and 2, GEMS, Rome.

FAO (1981a) Forest Resources of Tropical Asia, GEMS, Rome.

FAO (1981b) Los Recursos Forestales de la America Tropical, GEMS, Rome.

FAO (1981–1985), Monthly Bulletin, Tropical Forest Products in World Timber Trade, Rome.

FAO (1982), The state of the tropical forest, 2nd Expert Meeting UNEP, FAO, UNESCO, Rome.

FAO (1982), World Forest Products Demand and Supply 1990 and 2000, Rome.

FERGUSSON, K. (1969), Ghana Hardwoods, C. Tinling, Prescot, Lancashire, 109 pp.

FONTAINE, R. G. (1986), Management of humid tropical forests, Unasylva 38, 154, 16–21.

FREESE, F. (1962), Elementary Forest Sampling, Agriculture Handbook No. 232, U.S. Dept. of Agriculture, Forest Service, Washington, D.C., 91 pp.

GANGULY, J. K. (1963), Artificial regeneration of sal and teak, Indian Forester 89, 4, 272–274.

GILBERT, G. (1984), La masse forestière Congolaise, Bois et Forêts des Tropiques 204, 3–19.

GLERUM, B. (1962), Forest Inventory in the Amazon Valley, Report No. 1492, FAO, Rome.

GRIFFITH, A. L., BAKHSHI SANT RAM (1943), Yield and stand tables for sal (Shorea robusta) high forest, Indian Forest Records, Silviculture Series 4A, 4, Dehra Dun.

GROULEZ, J. (1963), L'okoumé dans le sud de son aire, Bois et Forêts des Tropiques 89, 37–42.

HADEN-GUEST, S. et al. (1956), A World Geography of Forest Resources, Ronald Press, New York, 736 pp.

HADLEY, M., LANLY, J. P. (1983), Tropical forest ecosystems: Identifying differences, seeking similarities, Nature and Resources 29, 1, 2–19.

HEINSDIJK, D. (1957), Forest Inventory in the Amazon Valley, FAO report No. 601, 135 pp.

HEINSDIJK, D. (1960), Interim Report to the Government of Brazil on the Dry Land Forests on the Tertiary and Quaternary South of the Amazon River, Report No. 1284, FAO, Rome.

HUFFEL, G. (1926), Économie forestière, Tom III, Paris.

JACOBS, M. R. (1979), Eucalypts for planting, FAO, Rome, 677 pp.

JANKO, J. (1947, 1948), How Statistics Form a Picture of the World and Life, Vols I and II, Union of the Czech Mathematicians and Physicians, Prague, 138 and 163 pp. (in Czech).

JEFFERS, J. N. R. (1960), Experimental Design and Analysis in Forest Research, Almquist and Wiksell, Stockholm–Uppsala, 157 pp.

JENÍK, J. (1973), Tropical rain forest, Živa 21, 1, 2–5; 2, 42–45; 3, 82–84; 4, 112–126 (in Czech).

JEYADEV, T. (1954), Working Plan for the Nilgiris Forest Division for the Period 1954 to 1964, Madras.

KAUL, P. N., FOTIDAR, A. N. (1961), Management of the forests of Jammu and Kashmir State, Indian Forester 87, 11, 667–677.

KORF, V. et al. (1972), Forest Mensuration, SZN, Prague, 327 pp. (in Czech).

LAMB, A. F. A. (1973), *Pinus caribaea,* Commonwealth Forestry Institute, Oxford, 254 pp.

LAMPRECHT, H. (1954), Über Strukturuntersuchungen im Tropenwald, Z. Weltforstwirt. **17**, 5, 162–168.

LAMPRECHT, H. (1972), Einige Strukturmerkmale natürlicher Tropenwaldtypen und ihre waldbauliche Bedeutung, Forstwiss. Cbl. **91**, 4–5, 270–277.

LANLY, J. P. (1982), Tropical forest resources, FAO Forestry Paper No. 30, Rome, 106 pp.

LANLY, J. P., CLEMENT, J. (1979), Present and future natural forest and plantation areas in the tropics, Unasylva **31**, 123, 12–20.

LAURIE, M. V., BAKSHI SANT RAM (1940), Yield and stand tables for teak plantations in India and Burma, Indian Forest Records, Silviculture Series **4A**, 1, Dehra Dun.

LE CACHEUX, P. (1955), Analyse statistique de la forêt tropicale en vus de son utilisation pour la production de la cellulose, J. Agric. Trop. Bot. Appl. 1–2, 1–17.

LONGWOOD, F. R. (1962), Present and Potential Commercial Timbers of the Caribbean, Agriculture Handbook, No. 207, U.S. Dept. of Agriculture, Forest Service, Washington, D.C., 167 pp.

LÖTSCH, F., HALLER, E., HENNING, N. (1967), Beitrag zur mathematischen Formulierung abnehmender Stammzahlverteilung, 14. Congress IUFRO, München, Sect. 25.

LÖTSCH, F., ZÖHRER, F., HALLER, K. E. (1973), Forest Inventory, Vol. 2, München–Basel–Wien.

MALAISSE, F. (1978) The Miombo ecosystem, in Tropical Forest Ecosystems, UNESCO–UNEP–FAO, Paris, pp. 589–606.

MALM, R. L. (1974), Ghana timber trends and prospects 1961–1972, Forest Products Research Institute, Kumasi.

MAYR, H. (1909), Waldbau auf naturgesetzlicher Grundlage, Berlin.

MACDONALD, T., MUJUNDAR, R. B. (1946), Working Plan for the Allapalli–Pedigundam Ranges, South Chanda Forest Division for the Years 1946–1961, Chanda.

McCOMBE, B., KRIEG, V. (1977), Properties of Laotian Timber Species, UNDP/UNIDO-ESCAP, Bangkok.

MENON, N. N. (1945), Working Plan for the Shencotta Forest Division, Konni.

MEYER, H. A. (1933), Eine mathematisch-statistische Untersuchung über den Aufbau des plenterwaldes, Schweiz. Z. Forstwesen **84**, 33–46, 88–103, 124–131.

MEYER, H. A. (1953), Forest Mensuration, Pennsylvania, 357 pp.

MYSLIVEC, V. (1957), Statistical Methods in Agricultural and Forest Research, SZN, Prague, 555 pp. (in Czech).

NAIR, C. T. S. et al. (1985), Intensive multiple-use forest management in the tropics, Forestry paper No. 55, FAO, Rome, 180 pp.

O'CONNOR, A. J. (1935), Forest research with special reference to planting distances and thinning, Br. Emp. For. Cong., South Africa.

PAQUET, J. et al. (1969), Reconnaissance Survey of Lowland Forest of Laos, The Canadian Agency for International Development, Vientiane.

PARRY, M. S. (1955), Simplified thinning schedules, Tang. For. Div. Silvicult. Circular No. 14.

PERSSON, R. (1983), Forestry in Laos, Paper for a conference, Swedish International Development Agency, Vientiane, 70 pp.

PIERLOT, R. (1968), Une technique d'étude de la forêt dense en vue de son aménagement: La distribution hyperbolique des grosseure, Bull. Soc. R. For. Belg. **2**, 122–130.

PIRES, J. M. (1978), The forest ecosystems of the Brazilian Amazon: Description, functioning and research needs, in Tropical Forest Ecosystems, UNESCO–UNEP–FAO, Paris, pp. 607–627.

PLOKHINSKII, N. A. (1937), Statistical Methods in Zootechnics, Selkhozgiz, Moscow (in Russian).

PRACNA, J. (1978), Problems of quality sortimentation of imported timber with regard to logging and transportation, Dissertation, Agricultural University, Prague (in Czech, unpublished).

PRIESOL, A. (1971), Development of a forest stand: Growth changes in time, in Bases of Growth and Production of Forests, M. Vyskot et al. (eds.), SZN, Prague, pp. 218–273.

PRINGLE, S. L. (1976), Tropical moist forests in world demand, supply and trade, Unasylva **28**, 112–113, 106–118.

PRINGLE, S. L. (1979), The outlook for tropical wood imports, Unasylva **31**, 125, 10–18.

PRYOR, L. D. (1959), Evolution in *Eucalyptus,* Austr. J. Sci. 1, 45–49.

PUDDEN, H. H. C. (1955), The pruning and early thinning of exotic softwoods in Kenya, For. Dept. Kenya, Pamphlet No. 13.

RAO, Y. S., CHANDRASEKHARAN, C. (1983), The state of forestry in Asia and the Pacific, Unasylva **35**, 140, 11–21.

RICHARDS, P. W. (1952), The Tropical Rain Forests: An Ecological Study, University Press, Cambridge, 450 pp.

RICHARDSON, S. D. (1969), Aspects of Forestry in Ghana, University of Wales.

ROLLET, B. (1974), L'Architecture des Forêts Dense Humides Sempervirents de Plaine, Centre Technique Forestier Tropical, Nogent sur Marne, 298 pp.

SAINT-AUBIN DE, G. (1963), La forêt du Gabon, Centre Technique Forestier Tropical, Nogent sur Marne, 208 pp.

SASSON, A. et al. (1978), Tropical Forest Ecosystems, UNESCO–UNEP–FAO, Paris, 683 pp.

SCHAEFFER, L. (1931), Sur trois modes des calcul de la possibilité des futaies jardinées, Nancy.

SCHIMPER, A. F. W. (1903), Plant Geography (Upon a physiological basis), Oxford.

SCHMIDT, R. (1987), Tropical rain forest management, Unasylva **39**, 156, 2–17.

SCHMITHÜSEN, F. (1976), Forest utilization contracts on public land in the tropics, Unasylva **28**, 112–113, 52–73.

SETH, S. K. et al. (1959), Yield and stand tables for plantation teak (*Tectona grandis* L. f.), Indian Forest Records, N. S., Silviculture, **9**, 4, 216 pp. Dehra Dun.

SETZER, O. (1986), Rapport sur l'aménagement dans l'Unité Forestière d'Aménagement No. 6, FAO–PRC 80/005, Brazzaville.

SEWANDONO, R. (1956), Southeast Asia, in A World Geography of Forest Resources, S. Haden-Guest et al. (eds.), Ronald Press, New York, pp. 491–517.

SINGH, B. P. (1957), Working Plan for the Haldwani Forest Division: Western Circle (Uttar Pradesh), Haldwani.

SINHA, J. N. (1961), Revised Working Plan for the Forests of Porahat Division, Patna, Bihar.

SOMMER, A. (1976), Attempt at an assussment of the world's tropical moist forests, Unasylva **28**, 112–113, 5–25.

STEFFERUD, A. et al. (1949), Trees, The Yearbook of Agriculture, U.S. Department of Agriculture, Washington, D.C., 944 pp.

STEINLIN, H. J. (1982), Monitoring the world's tropical forests, Unasylva **34**, 137, 2–9.

STREETS, R. J. (1962), Exotic Forest Trees in the British Commenwealth, Clarendon Press, Oxford, 759 pp.

SVOBODA, P. (1952), Life of the Forest, SZN, Prague, 895 pp. (in Czech).

TAYLOR, C. J. (1962), Tropical Forestry with Particular Reference to West Africa, Oxford University Press, London, 166 pp.

THOMPSON, R. (1973), More Lesser Known Commercial Timber Trees of Ghana, Ghana Timber Marketing Board, Takoradi, 11 pp.

TROUP, R. S. (1921), The Silviculture of Indian Trees, Vols I–III, Clarendon Press, Oxford.

TURNBULL, K. J. (1963), Population dynamics in mixed forest stands, Dissertation, University of Washington, 196 pp.

UNCTAD (1982), Statistics of Exports and Imports of Tropical Timber, Geneva.

VÁCLAV, E. (1983), Forest Plantation of Equatorial and Southern Africa, Agricultural University, Prague, 110 pp. (in Czech).

VAZQUEZ SOTO, J. (1962), General considerations on Mexican conifers, in Seminar and Study Tour of Latin-American Conifers, Instituto Nacional de Investigaciones Forestales, Mexico, pp. 1–119.

VIDAL, J. (1956), La Vegetation du Laos, 1. Conditions Ecologiques, Toulouse.

VYSKOT, M. et al. (1971), Bases of Growth and Production of Forests, SZN, Prague, 440 pp. (in Czech).

WALTER, H. (1962), Die Vegetation der Erde, Vol. 1, Die Tropischen und Subtropischen Zonen, VEB
G. Fischer Verlag, Jena, 538 pp.

WANGH, CHI-WU (1961), The Forests of China, M. Moors Cabot Foundation, Publ. No. 5, Harvard
University, Cambridge, Mass., 313 pp.

WECK, J. (1957), Die Wälder der Erde, Springer-Verlag, Berlin–Göttingen–Heidelberg, 152 pp.

WECK, J., WIEBECKE, C. (1961), Weltforstwirtschaft und Deutschlands Forst- und Holzwirtschaft, BLV,
München.

WHITMORE, T. C. (1975), Tropical Rain Forest of the Far East, Clarendon Press, Oxford, 282 pp.

WILLAN, R. L. (1964), Thinnings experiments in exotic softwoods, Tang. For. Div. Tech. Note (Silvic.)
No. 63.

WORMALD, T. J. (1975), *Pinus patula,* Commonwealth Forestry Institute, Oxford, 199 pp.

ZÖHRER, F. (1969), The application of the beta function for best fit of stem-diameter distributions in
inventories of tropical forests, Tagungsbericht der Arbeitsgruppe der Sektion 25, IUFRO, Reinbek.

Cahier des statistiques forestières, Ministère de l'Economie Forestière, Brazzaville, 1987.

Décret No. 84/910 du 19/10/1984 portant application du Code forestièr, Brazzaville, 1984.

External Trade Statistics of Ghana, Vol. 22, No. 12, Central Bureau of Statistics, Accra, 1972, 258 pp.

Ghana Annual Reports of the Forestry Department, Ministry of Lands and Mineral Resources, Accra,
1970–1974.

Index of tree species Latin names

K. grandifoliola 50, 53, 65
K. ivorensis 39, 48, 50, 53, 60–62, 65, 113
K. nyasica 167
K. senegalensis 52
Kigelia aethiopica 52
Klainedoxa gabonensis 75, 80, 96, 112, 114
K. gabonensis, var. *oblongifolia* 53
Koompasia excelsa 13

Lagerstroemia 128, 130, 139–141, 147, 154
L. angustifolia 21
L. calyculata 145
L. floribunda 142
L. lanceolata 181
L. parviflora 191
L. speciosa 145
Laguncularia 52
Lannea 24
L. schimperi 169
Larix decidua 5
L. gmelini 5, 8
L. laricina 5
L. occidentalis 5
L. sibirica 5
Libocedrus 17
L. decurrens 8
Liquidambar 12
L. styraciflua 8
Litsea polyantha 130
Lophira 50
L. alata 17, 34, 48, 50, 53, 55, 57, 114
L. lanceolata 52
Lovoa 17, 79
L. trichilioides 39, 50, 53, 54, 71, 78, 97, 113

Macaranga 63, 71, 76
Machilus 131, 141
Maesopsis eminii 53, 54
Mallotus philippinensis 194
Mangifera indica 52
Manglietia glauca 131
Manilkara bidentata 19
M. lacera 53
Mansonia altissima 39, 50, 53, 55, 57, 61, 65
Melaleuca leucadendron 21
Melanorrhoea laccifera 130, 141
Melia 131, 141
Mesua ferrea 130
Michelia 9
M. champaca 191
Microberlinia brazzavillensis 79

Millettia laurentii 79, 98
M. stuhlmannii 178
Mimosa nigra 52
Mitragyna 48, 55
M. ciliata 39, 53, 76, 79, 114
M. stipulosa 53
Monodora myristica 50
Monopetalanthus 112
M. heitzii 114
M. letestui 101
M. pellegrini 80
Monotes 167
Musanga 63, 76
M. cecropioides 71, 76
Mussaenda chippii 49
Myrianthus 50
Myrica 12
Myristica athenuata 154

Nauclea diderrichii 48, 53, 56, 75, 76, 79, 98, 120
N. trillesii 114
Nesogordonia papaverifera 39, 50, 53, 55, 57, 61, 65, 114
Newtonia leucocarpa 114
Nothofagus 9

Ochroma lagopus 35
O. pyramidale 19
Ocotea cymbarum 35
O. rodiaei 19, 35
O. usambarensis 20, 167
Olea europaea 9
O. welwitschii 167
Ongokea gore 53
Ormosia cambodiana 146
Ostryoderris 167
Oxystigma oxyphyllum 76, 79, 97, 114
Oxytheutera 129

Pahudia cochinchinensis 132
Pandanus 76
Paraberlinia bifoliolata 112, 114
Parashorea 17, 33, 34, 38, 154
P. lucida 133
P. stellata 128, 131, 133
Parinari curatellifolia 169
P. excelsa 53
P. glabra 101
Parkia 21, 52
P. bicolor 48, 50
Parrotia jaquemontiana 210

T. *tomentosa* 128, 131, 181, 185, 191

Tessmania lescrauwaetii 96, 101

Testulea gabonensis 79, 114

Thuja occidentalis 5

T. plicata 5

Tilia americana 5

Toona febrifuga 131

Trema 63

Triplochiton 50

T. scleroxylon 20, 33, 34, 38, 39, 41, 50, 53, 54, 57, 61, 65, 97

Tsuga canadensis 5

T. heterophylla 5

Turraeanthus africana 17, 50, 53, 54, 57, 79

Uapaca 14, 17, 21, 53, 71, 72, 76, 167

U. guineensis 80, 101

U. kirkiana 169

U. nitida 169

U. pilosa 169

Ulmus americana 5, 8

U. glabra 8

U. rubra 8

U. villosa 210

Vateria indica 154

Vatica 128

V. bancana 21

V. dyeri 129, 130, 141, 142, 154

V. roxburghiana 130

Vernonia 71, 76

Viburnum nervosum 210

Virola 33, 35, 38

V. surinamensis 19

Vitex 21, 53, 170

V. cuneata 52

V. pachyphylla 114

V. pubescens 131, 141

Vochysia maxima 152

Vouacapoua americana 152

Xeroderris 167

X. stuhlmannii 169

Xylia 21

X. kerrii 128, 131, 135, 142

X. xylocarpa 130, 191

Xylopia 152

X. odoratissima 169

Zizyphus 26

Index of tree species trade or vernacular names

celtis 65
champak 130, 131
cherry mahogany 53
chuglam 131
cocoa tree 46, 68, 110
coconut tree 46
coffee tree 68, 110

dabema 53, 55, 59–61, 65, 79–83, 89, 94, 114, 115, 119
dahoma 53, 55
danta 53, 55, 61, 65
dedwar 210
deodar 210, 211, 213, 214
diambi 53, 60, 80
diania 79, 89, 94
dibetou 53, 67, 78, 81, 83, 88, 97, 98, 102, 106, 113, 115
dina 79
dipterocarp 139, 143, 151
douka 79, 97, 98, 106, 113, 115, 118, 123
doussié 56, 79, 81, 98, 102, 106, 113, 115

East African cedar 215
eba 89
ebana 114, 118
ebiara 79, 80, 94, 98, 106, 114, 115, 119
ebonies 98, 132, 133
edinam 55
edoun 114, 116
ekki 34, 53, 55
ekop 114, 115, 119
ekoune 79–83, 94, 114–119
ekumba-apio 80
emri 53, 54
emien 79, 80, 114, 115
esa 53
espavel 35
essessang 54, 79, 89, 94, 114, 115
essia 53, 80, 89, 98
eteng 115
eucalypts 9, 12, 73, 107, 120, 167, 195, 196, 198, 199, 201, 202, 207, 208
eveuss 80, 89, 114–116, 119
evino 114

faro 79, 98, 109, 114, 115, 119
fir 5, 11, 44
framiré 54, 65
fromager 54, 114, 115

gheombi 116, 119
gombé 114
greenheart 35
guarea 53, 55, 59, 61, 79
guarea black 53
guarea scented 53
guarea white 53

haldu 131
hevea 46
hickory 55
Himalayan cedar 210

idigbo 53
ilomba 53, 54, 73, 79–83, 94, 97, 102, 113, 115, 117–119, 122, 123
imbuia 38
Indian laurel 131
Indian satinwood 56
iroko 33, 34, 52, 55, 63–65, 67, 79, 88, 97, 98, 101, 106, 109, 113
ironwood red 34, 53, 55
izombé 79, 106, 114, 115, 118

kabok 131
kaku 55
kanda 109
kapok 54
kapur 34, 38
kedondong 131
kelu 210
keruing 34, 38, 130, 133
kevazingo 98, 114–116, 118
khaya 54, 59–61, 65, 67, 73, 106, 119
khaya bigleaf 53
khaya red 53
kilar 210
kokrodua 53, 55, 62–65, 67, 98, 106
kosipo 53, 55, 65, 67, 73, 79, 81, 97, 98, 102, 106, 113, 115, 119
kotibé 48, 55, 56, 60–62, 106, 114, 115
koto 41, 55
krabak 130–132
kra-thon 131
kusia 56
kyere 55

larch 5
lati 79
lauan 33, 38, 49
limba 33, 34, 41, 53, 73, 79, 81, 88, 97, 98, 101,